SERIES ON SEMICONDUCTOR
SCIENCE AND TECHNOLOGY

Series Editors

A. N. Broers C. Hilsum R. A. Stradling

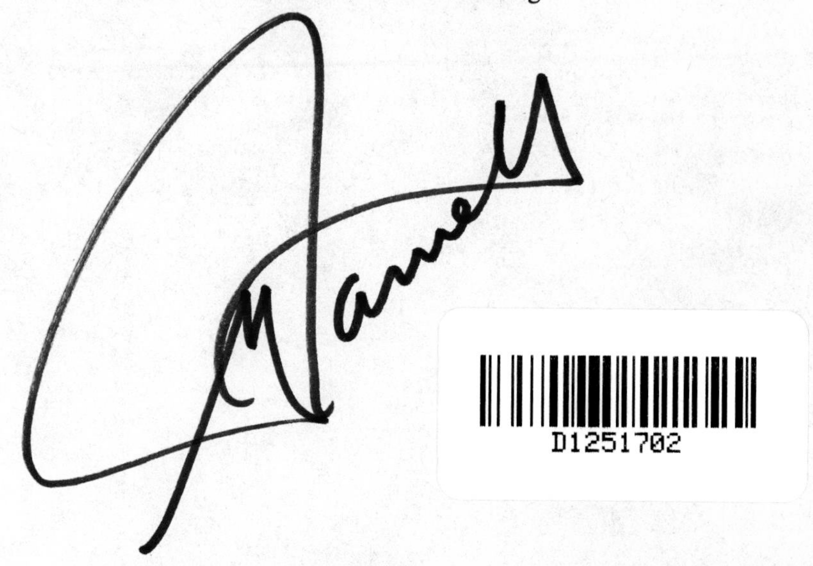

SERIES ON SEMICONDUCTOR SCIENCE AND TECHNOLOGY

1. M. Jaros: *Physics and applications of semiconductor microstructures*

Physics and Applications of Semiconductor Microstructures

M. Jaros
Department of Physics
The University, Newcastle upon Tyne

CLARENDON PRESS · OXFORD
1989

Oxford University Press, Walton Street, Oxford OX2 6DP
Oxford New York Toronto
Delhi Bombay Calcutta Madras Karachi
Petaling Jaya Singapore Hong Kong Tokyo
Nairobi Dar es Salaam Cape Town
Melbourne Auckland
and associated companies in
Berlin Ibadan

Oxford is a trade mark of Oxford University Press

Published in the United States
by Oxford University Press, New York

British Library Cataloguing in Publication Data
Jaros, M.
Physics and applications of semiconductor microstructures.
1. Semiconductors. Physical properties
I. Title
537.6'22
ISBN 0-19-851994-X
ISBN 0-19-853927-4 (Pbk.)

Library of Congress Cataloging in Publication Data
Jaros, M.
Physics and applications of semiconductor microstructures/M. Jaros.
p. cm.—(Series on semiconductor science and technology: 1)
Bibliograhy: p. Includes index.
1. Semiconductors. 2. Microstructure. 3. Quantum wells.
4. Superlattices as materials. 5. Microelectronics. I. Title.
II. Series.
QC611.J27 1989 537.6'22—dc19 88-22561
ISBN 0-19-851994-X
ISBN 0-19-853927-4 (Pbk.)

Typeset by The Universities Press (Belfast) Ltd
Printed in Great Britain
by Biddles Ltd.
Guildford & King's Lynn

Preface

This book is an extended version of my lecture notes for a course I have given under a similar title to third-year undergraduate and first-year postgraduate students in physics and microelectronics. Its aim is to outline basic concepts that account for new physical phenomena brought into existence by the reduced size and dimensionality characteristic of novel semiconductor structures. These concepts are developed as a natural extension of the ideas normally used to describe electronic structure and physical properties of macroscopic crystalline solids and semiconductor devices. A substantial portion of the book is devoted to representative examples of the application of such concepts in electronic and optical devices.

The characteristic dimension of the heterojunction structures this book is really about is of order 10 nanometres (10^{-8} m). Therefore, strictly speaking, they should be called nanostructures. However, I am confident that most of us ordinary mortals take 'micro' to mean 'very small' and it is in this common sense that the term microstucture is used throughout the book.

There are a number of excellent undergraduate textbooks covering standard topics of solid-state physics and semiconductor devices at a variety of levels of difficulty. However, as far as I am aware, only at a very advanced level are the physics and applications of heterojunction semiconductor structures covered in the existing literature. This is in sharp contrast with the enormous interest this field has been enjoying in recent years, and with the amount of current research and training effort, both in industry and in academia. This book shows that the basic physics and device concepts concerning semiconductor microstructures are as accessible as the physics of macroscopic semiconductor systems. I have endeavoured to avoid unnecessary repetition of the material in standard textbooks. However, there is a certain degree of overlap in the presentation here of the fresh material with that of the conventional semiconductor physics, since new ideas are introduced by developing them from the familiar ones. It also enables me to present some background topics from a rather more quantitative angle than is customary in textbooks on semiconductor devices. This is, I believe, achieved at a level of difficulty that is still well within the limits acceptable in undergraduate (physics or engineering) courses. I find it particularly gratifying that I have been able to translate even the most difficult concept of heterojunction band offset (which is of key importance in the present context) into the simplest language of nearly-free-electron theory, without having to sacrifice the semi-quantitative approach adopted in other parts of this book or the clarity expected of an elementary text.

Bearing in mind that the level of study spans the final years of a degree course or the first year of a postgraduate course, I have assumed that the reader is familiar with the elementary facts of atomic and semiconductor physics. However, I recognize that the teaching of quantum concepts is uneven and often badly neglected in engineering and materials science departments, where most of my potential readers probably come from. I have therefore included (in the Appendices and in several places in the text) a summary of the relevant elementary passages from quantum mechanics. In the introductory Chapters 1 to 5—which are very brief but self-contained—the reader will find all the necessary concepts and mathematical expressions required in the text that follows. Hence, these chapters serve a dual purpose, as a summary/reminder of the background (conventional) solid-state physics, and as a set of general equations that are invoked (but not repeated) in the subsequent chapters, where they are used in a different physical context.

All formulae derived in the text are presented in SI units. However, numerical examples supporting theory are given in units normally used in the technical literature. I believe that the reader should be aware that, for example, the lattice constant is invariably quoted in ångstroms and not in metres, the work function of a metal in electron-volts and not in joules, etc. A table of fundamental constants with the relevant conversion factors is included at the back. The illustrations used to support the text with examples of empirical data or compilations of parameters are taken from the specialist literature and simplified by stripping them of details of secondary importance that would be difficult to explain at the level of sophistication chosen in this book. Accordingly, the source is immaterial and to avoid any misunderstanding no attempt is made to refer to original literature.

The field of microstructure physics and applications is highly multidisciplinary. This means that a large number of variables are mentioned that, in the fields where they originated, are known under the same letter. For instance, n is invariably used to label the refractive index in optics, the electron density in electron transport, the band index in theory of electronic structure, the principal quantum number in atomic physics, etc. Rather than invent new labels, I have retained the usual symbol whenever possible and, if necessary, redefined it in the new context to avoid any confusion. I have endeavoured to keep the notation as simple as possible. For example, when dealing with a one-dimensional model, I drop the vector symbolism.

The microstructure physics and devices chosen for discussion are by no means an exhaustive survey of what is available in specialized literature. Such a survey would be premature at this introductory level of presentation, since most of the new devices in this field have not been implemented outside research laboratories and probably never will be in their present form. I have chosen to discuss only the archetypal effects and device ideas that will no doubt survive the test of time in one form or another. Even so, the book covers a large number of very disparate topics (e.g. crystal growth, lithography, hot-electron devices, etc.), each of which could alone provide more than enough material for a book at any level of difficulty. I am hoping that—given a minimum background in solid state physics and devices—I tell the reader just enough to hold his attention throughout the text, and to provide him with basic orientation in microstructure physics and application and with a starting point for specialized reading. This pragmatism creates a certain degree of imbalance, in that topics that are normally well covered in elementary textbooks, and where the relevant techniques established for macroscopic structures can be made use of, appear to be rather neglected (e.g. magneto-transport, Bloch oscillations). On the other hand, certain more unusual material required for novel device applications receives a disproportionate amount of attention (e.g. higher order susceptibilities). Furthermore, some of the intellectually most stimulating topics are also among the most demanding conceptually, and their coverage here is of necessity very brief (e.g. quantum Hall effect, superconducting junctions, non-linear response, etc.).

The bibliography is organized so as to help the uninitiated to choose the relevant text according to the level of difficulty. Advanced texts recommended there cover all topics touched upon in the text. Naturally, I list only a small fraction of the available literature and the reader should know that there are many other equally desirable texts.

The problems at the end of each chapter are intended to encourage readers to 'design' their own microstructures for given purposes and to appreciate the power as well as the limitations of the mathematics presented in the text. Solutions to these problems are outlined at the end.

At the end of the book, the reader will also find the Periodic Table of the Elements. I hope that after reading the book he will no longer see this table simply as a record of the number of electrons and protons per atomic species. Instead, it should always remind him of the number of materials and structures available to us for experimenting with new physics and devices. Thanks to the recent advances in materials science, the opportunities open to us in this field are limited only by our imagination.

April 1988 M. J.

Contents

1

Description of non-interacting electrons in periodic structures: free-electron model

Crystals are systems consisting of a large number of regularly positioned atoms. In order to describe the physical properties of a given crystal, we must understand how the behaviour of a single isolated atom is modified when the atoms are joined together to form a crystalline lattice. Then we can set up a model that will enable us to describe the basic concepts peculiar to the crystalline system in question. We shall begin with the free-electon model, which is often used to describe elementary properties of metals such as conductivity and which is also the simplest model of a solid in general.

Let us consider, for example, a free sodium (Na) atom and think about what will happen if we decide to joint such atoms together to form sodium metal. Just like any other atom (see Appendix 1) the sodium atom consists of a nucleus surrounded by electrons orbiting in quantum states. In sodium these states are described as 1s, 2s, 2p, and 3s. Here the first letter is the value of the principal quantum number n ($n = 1$, 2, and 3) and the second refers to the angular momentum quantum number l($l = 0$ is labelled as s, $l = 1$ as p, and $l = 2$ as d). Although we do not need to invoke the quantum-mechanical description of atoms in any detail here, it is well worth reminding ourselves of this language of quantum mechanics; it may add a deeper meaning to some of our future discussions.

The ionization energies required to lift an electron from a quantum state and remove it from the sodium atom are shown in Fig. 1.1. It is an established practice to use electron-volts (eV) as a unit of energy. 1 eV equals 1.602×10^{-19} joules. The energies of bound states are negative relative to the vacuum. For example, the ground (lowest) state of a free hydrogen atom is at -13.6 eV. The unit Rydberg or simply Ry, (1 Ry = 13.6 eV) is often used in the literature.

With the exception of the outer (3s) state, which is singly occupied, all other quantum states are fully occupied and form 'closed shells'. These

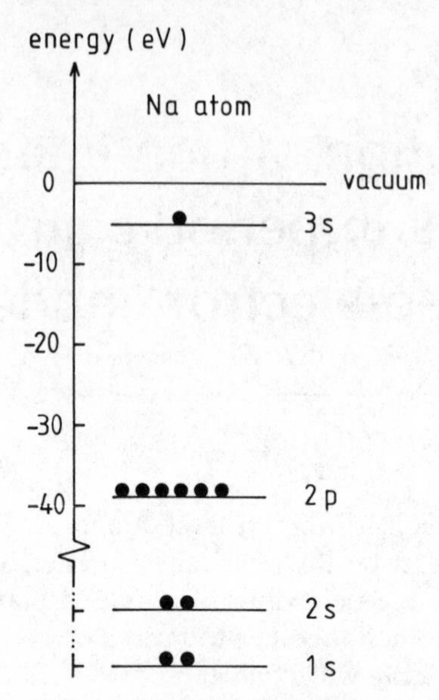

Fig. 1.1. The energy levels of 1s, 2s, 2p, and 3s electrons at an isolated sodium atom (in electron-volts).

'core' electrons orbit very close to the nucleus compared to the outer 'valence' electron whose orbit is more extended. It is this valence electron that is of particular interest to us.

In quantum mechanics, we compute the electron energy levels and wave functions from the Schrödinger equation. In this equation, we must represent the attractive potential due to the positively charged nucleus, and the repulsive potential due to the electron–electron interaction. The replusive contribution cancels a substantial part of the attractive Coulomb potential. The net potential experienced by a valence electron in the region of space outside the range of the core electron orbit is given by elementary electrostatics. To a test charge (an electron) positioned at large distances from the nucleus, the core electrons look like point charges that shield the nucleus, so that the outside electron sees a potential due to only one (positive) point charge at the Na nucleus. This is illustrated schematically in Fig. 1.2. In the vicinity of the nucleus, the potential experienced by our 3s electron is not that of a simple point charge. This is indicated by a broken line in Fig. 1.2. If it were, the energy of the 3s state would be -13.6 eV, i.e. that of an electron in a free hydrogen atom. In fact, we know from Fig. 1.1 that the energy is only about -5 eV.

Let us now consider what happens when sodium atoms are brought together to form a crystalline lattice. If the separation between nearest-

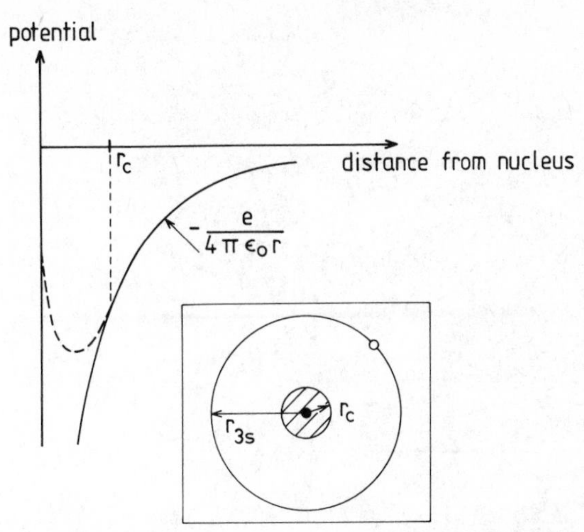

Fig. 1.2. The potential seen by a valence electron of sodium as a function of the distance between this electron and the nucleus; r_c is the core radius marking the volume near the nucleus where the core electrons (1s, 2s, and 2p) are localized. The inset shows a sketch of a 3s valence electron orbiting around the nucleus shielded by the core electrons.

neighbour atoms is small enough, we end up with a system in which the valence electron wave functions strongly overlap. A typical nearest-neighbour separation in a solid is 2–4 Å (1 Å $= 10^{-10}$ m) so that the core electrons, whose orbits have radii smaller than 1 Å, do not overlap with their counterparts on the neighbour atoms. These electrons retain their atomic character even when the atom becomes part of a solid, and do not affect the physical properties of crystals. We shall simply ignore them in our future considerations.

The total valence charge density of a sodium crystal can now be viewed as a superposition of overlapping atomic contributions, i.e. orbits such as those in the inset of Fig. 1.2. This is illustrated in Fig. 1.3. We can see that if the interatomic separation is small enough, the total valence charge density becomes featureless and flat. Since most metals form BCC or FCC cubic crystals (Fig. 1.4), each atom will have typically eight equivalent neighbours, so that the angular variation of the charge density will also be weak, i.e. the total charge density curve will be flat irrespective of the direction in the lattice we choose to follow. It is borne in mind that under these circumstances it is no longer profitable to think of the valence charge density in terms of a superposition of contributions from each individual atom, since the atomic signature is reduced to the mere fact that (in the case of sodium) each atom supplies one electron. This also means that the relevant part of the total crystal potential is

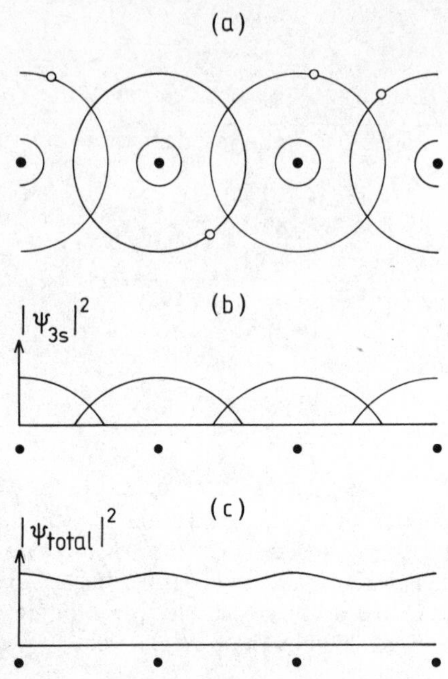

Fig. 1.3. (a) Sodium atoms, with their valence electron orbits shown explicitly as in Fig. 1.2, are joined to form a one-dimensional lattice. (b) A sketch of the overlapping charge of the atomic valence electrons. (c) When the individual atomic contributions shown in (b) are added up to form the total charge density distribution in the crystal, one obtains a featureless curve.

featureless and that its details can be ignored. That, however, implies that, instead of writing the total (crystal) potential V_c as a linear combination of atomic potentials such as that shown in Fig. 1.2, we can approximate V_c by a constant. Thus, the quantum-mechanical description of an electron in a simple metal (e.g. sodium) reduces to solving the Schrödinger equation for a free electron,

$$-\frac{\hbar^2}{2m}\left(\frac{\partial^2}{\partial x^2}+\frac{\partial^2}{\partial y^2}+\frac{\partial^2}{\partial z^2}\right)\psi = E\psi. \tag{1.1}$$

This corresponds to a classical description of a free particle whose energy is $E = p^2/(2m)$, where p is the linear momentum and m is the electron mass. In quantum mechanics, we replace p with an 'operator' such that $\hat{p}_x = -i\hbar\partial/\partial x$, $p_y = -i\hbar\,\partial/\partial_y$, etc., where $\hbar$ is the Planck constant divided by 2π. If we use the de Broglie relation $\mathbf{p} = \hbar\mathbf{k}$, where $\mathbf{k}$ is the wave vector ($|\mathbf{k}| = 2\pi/\lambda$ where λ is the electron wavelength: which means we attach to an electron both wave-like and particle-like properties), we

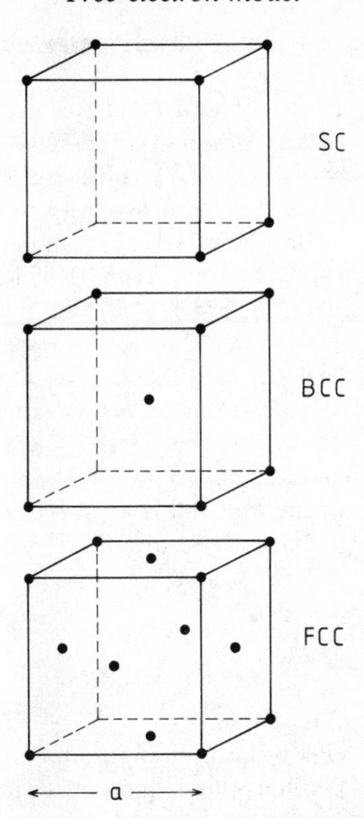

Fig. 1.4. The conventional unit cell (a is the lattice constant) of the simple cubic (SC), body-centered cubic (BCC), and face-centered cubic (FCC) lattices.

obtain for the energy $E = p^2/(2m) = \hbar^2k^2/(2m)$. But this is exactly what we obtain from eqn (1.1) if we use for the wave function $\psi = \text{const.} \times \exp(i\mathbf{k} \cdot \mathbf{r})$ where $\mathbf{r}$ is a position vector in Cartesian coordinates (x, y, z). This can be verified by substituting in eqn (1.1) for ψ and differentiating; in other words, the wave function of an electron in a simple metal is a plane wave and the corresponding charge density is $e\psi\psi^* \sim e \exp(i\mathbf{k} \cdot \mathbf{r}) \exp(-i\mathbf{k} \cdot \mathbf{r}) = e = \text{constant}$, as expected from our considerations above.

Clearly, the picture we have just arrived at is an approximation that holds good only for certain classes of metals. In other solids, and particularly in semiconductors and insulators, the interaction between the electrons and ions cannot be neglected and the charge density exhibits distinct features reflecting the atomic signature of the crystal potential. It is these interactions that we shall have to account for in the following chapters. However, in many instances of interest to us (e.g. semiconductors such as silicon), the electron interaction with the lattice of ions is weak and the kinetic energy terms described by eqn (1.1) dominate the solutions of the Schrödinger equation.

In the free-electron model of eqn (1.1) the electrons are 'free', i.e. independent of their atomic origin, and behave as an electron gas. However, the solution of eqn (1.1) must be specified by defining boundary conditions. This is a general property of differential equations and is not peculiar to our problem. However, the boundary conditions in our case are very special since the metal can be divided into identical cells and consequently the wave function, i.e. the solution of eqn (1.1), must also be the same in each cell. This is illustrated in Fig. 1.5 for a one-dimensional example. We can choose the length of our unit cell $L(L = ma$, where a is the nearest-neighbour atomic separation and m is a natural number) and demand that $\psi(x) = \psi(x + L)$. In a three-dimensional case the same must apply in all direction, i.e. $\psi(x, y, z) = \psi(x, y + L, z)$, etc. The unit cell is a cube of volume $V = L^3$. Since $\psi \sim \exp(i\mathbf{k} \cdot \mathbf{r})$, it is clear that only solutions with components k_x, k_y and k_z such that

$$k_x = \frac{2\pi}{L} n_x; \qquad k_y = \frac{2\pi}{L} n_y; \qquad k_z = \frac{2\pi}{L} n_z;$$

$$n_x, n_y, n_z = 0, 1, 2, \ldots \qquad (1.2)$$

i.e. the solutions that satisfy the periodic boundary condition, are valid solutions of eqn (1.1) when this equation describes electrons in a crystalline solid. It follows that any valid solution of eqn (1.1) can be identified in terms of three natural numbers, n_x, n_y, and n_z such that $E = \hbar^2 k^2/(2m) = E(n_x, n_y, n_z)$. For example, the lowest-energy solution of eqn (1.1) is given by $n_x = n_y = n_z = 0$. The next solution has $n_x = 1$, $n_y = 0$, $n_z = 0$, and so on. Let us now suppose that there are N valence electrons in volume V (concentration $n = NV^{-1}$). Many years ago, the so-called Pauli, or exclusion, principle was established, which states that only two electrons can 'share' any particular solution of the Schrödinger equation (the difference between them being only the orientation of their spin). If we now want to assign each of our N electrons a certain solution

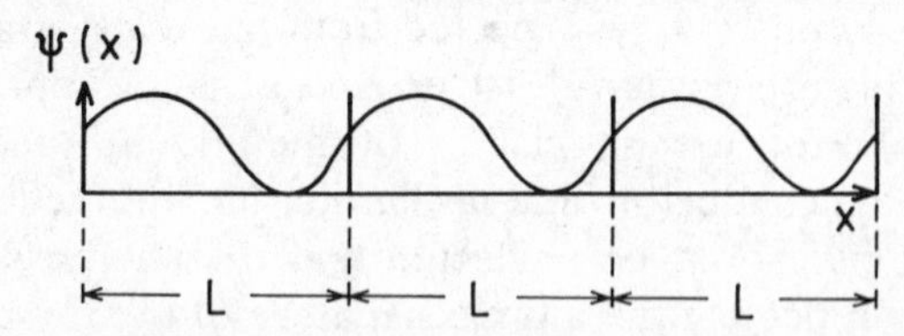

Fig. 1.5. The wave function of a one-dimensional crystal whose unit cell is of length L. It is a periodic function, i.e., it satisfies the periodic boundary conditions imposed by the crystal symmetry.

of eqn (1.1), we must reserve one set of numbers n_x, n_y, n_z (i.e. k_x, k_y, k_z) for two electrons. We can visualize this process as one consisting of placing two electrons at each point of a grid defined by n_x, n_y, n_z (or k_x, k_y and k_z). Since we start at zero energy ($k_x = k_y = k_z = 0$) and move upwards, and since energy $E \sim k^2$, i.e. it depends only on the length of the wave vector, the occupied k-points fill a sphere ($N \gg 1$) of radius k_{max}, where k_{max} is the length required to accommodate all N electrons. The volume in k space occupied by two electrons (i.e. the volume per one allowed state defined by eqn (1.2)) is $(2\pi/L)^3$. We can then express N as

$$N = 2\left(\frac{4\pi}{3} k_{max}^3\right)\left(\frac{L}{2\pi}\right)^3. \tag{1.3}$$

It is now possible to relate k_{max} to electron concentration $n = N/L^3$, i.e.

$$k_{max} = (3\pi^2 n)^{1/3} \tag{1.4}$$

The significance of this result is apparent from Fig. 1.6. We plot the electron energy $E = \hbar^2 k^2/(2m)$ versus k. We know that the allowed states form a ladder of levels (it is a discrete set). At zero temperature, i.e. when there is no external source of energy, the lowest $N/2$ states (with $k \leqslant k_{max}$) are occupied. If we know n, we can compute k_{max} (and consequently $E_{max} = \hbar^2 k_{max}^2/(2m)$) for any crystal. For example, in the case of sodium, we showed that each atom supplies exactly one valence electron, so that n is equal to the density of atoms. Sodium has a

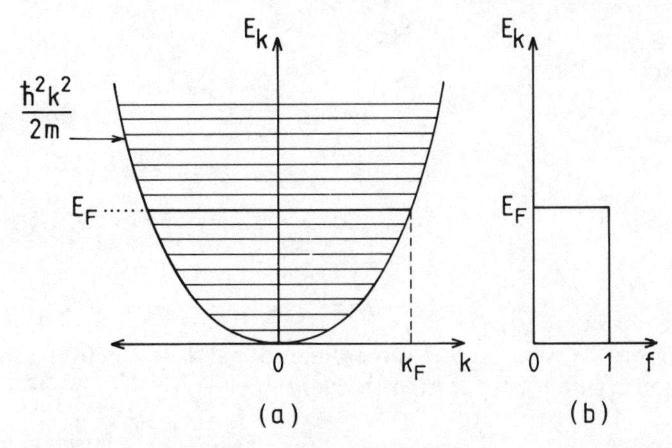

Fig. 1.6. (a) The free-electron energy $E = \hbar^2 k^2/(2m)$ as a function of k. The states below the Fermi energy E_F are doubly occupied and the states above E_F are empty (at zero temperature). (b) f is the probability of finding a state at energy E_k occupied (at zero temperature).

body-centred cubic lattice at 0 K. The length of the conventional unit cube is 4.23 Å (the lattice constant a, Fig. 1.4). There are two atoms (and therefore two valence electrons) per conventional cube so that $n = 2.65 \times 10^{22}\,\text{cm}^{-3}$, $k_{\text{max}} = 0.92 \times 10^8\,\text{cm}^{-1}$ and $E_{\text{max}} = 3.24\,\text{eV}$. Hence, all states with energy larger than 3.24 eV are empty. The energy and wave-vector of the uppermost occupied state (E_{max} and k_{max}, respectively) are normally referred to as the Fermi energy and wave-vector, E_F and k_F, respectively.

Finally, we can ask how many states are available at a given energy, for instance at E_F. We shall need this information when we come to consider the electron transport and other observables. The number of states dN per energy interval dE is the density of states $\rho = \mathrm{d}N/\mathrm{d}E$. In our model, the volume per allowed state in k space is $(2\pi/L)^3$. Thus, in a unit volume there are $(2\pi)^{-3}$ states. The number of electrons dN in volume $4\pi k^2\,\mathrm{d}k$ associated with interval dk is therefore

$$\mathrm{d}N = 2\frac{4\pi k^2\,\mathrm{d}k}{(2\pi)^3} \tag{1.5}$$

(the factor of 2 again means that there are two slots per state—the Pauli principle). We want dN/dE, so that we must express dk in terms of dE. We know that

$$E \equiv E_k = \frac{\hbar^2 k^2}{2m}, \tag{1.6}$$

so that

$$k = \left(\frac{2m}{\hbar^2}E_k\right)^{1/2} \tag{1.7}$$

and

$$\frac{\mathrm{d}k}{\mathrm{d}E} = \left(\frac{2m}{\hbar^2}\right)^{1/2}\frac{E_k^{-1/2}}{2}. \tag{1.8}$$

Substituting into eqn (1.5) for k and dk and dividing by dE gives

$$\rho = \frac{\mathrm{d}N}{\mathrm{d}E} = \frac{1}{2\pi^2}\left(\frac{2m}{\hbar^2}\right)^{3/2} E_k^{1/2}. \tag{1.9}$$

The density of states of a three-dimensional free-electron metal is a continuous (parabolic) function of energy E_k.

Summary

In a free-electron crystalline metal the quantum states satisfying the Schrödinger equation are simple plane waves. Each state is characterized by three constants n_x, n_y, and n_z such that the corresponding wave vector $\mathbf{k}$ satisfies the periodic boundary condition. The largest value of $\mathbf{k}$ associated with an occupied state at 0 K is the Fermi wave-vector $k_F = (3\pi^2 n)^{1/3}$, where n is electron density. The corresponding Fermi energy is $E_F = \hbar^2 k_F^2/(2m)$ and the density of electron states per unit volume is $\rho = (2m/\hbar^2)^{3/2} E_k^{1/2}/(2\pi^2)$.

Problems

1.1. Calculate the volume per atom and the free-electron Fermi energy E_F for lithium, sodium, and potassium, and then evaluate the density of states for each crystal at energy $E = E_F$.

1.2. Find an expression for the density of states of a two-dimensional metal.

1.3. What is the crystalline structure (e.g. BCC, FCC, etc.) of NaCl, Si, GaAs, and Na. What is the number of nearest-neighbour atoms in these crystals? What is the lattice constant and the nearest-neighbour separation (in angstroms)?

1.4. Use the procedure adopted in Problem 1.1 to calculate E_F in electron-volts for a silicon crystal.

1.5. Find k_F in Problem 1.1.

2
Quantum states of electrons interacting with a weak periodic potential: nearly-free-electron band structure

We must now ask what would happen if there were at least some weak interaction between the sea of free electrons and the ions sitting at the lattice sites of our model solid. The simplest way to visualize this interaction is to assume that the interaction is elastic, i.e. that electrons colliding with the lattice of ions do not lose energy but merely change their directions. Since the lattice is always a perfectly regular array of identical objects (atoms or groups of atoms) and electrons are represented by plane waves, the interaction is exactly analogous to the interaction of an optic ray with a grid and must lead to a diffraction pattern. Such a pattern is described by the Bragg law. This problem is discussed at length in practically all textbooks on solid-state physics, and a detailed knowledge of crystal structures and crystallographic notation is normally required. Our aim here is to identify basic concepts, and these can be demonstrated on simple structures and models. We shall, therefore, rely here on a common-sense grasp of crystal symmetry. Take, for example, a square lattice with a nearest-neighbour separation d (Fig. 2.1). Any lattice point can be defined by a (two-dimensional) position vector $\mathbf{R} = n\mathbf{a}_1 + m\mathbf{a}_2$, where n and m are integers. A free electron interacting with the lattice is represented by a plane wave with wave-vector $k = 2\pi\hat{\mathbf{u}}/\lambda$, where $\hat{\mathbf{u}}$ is a unit vector and λ is wavelength. For the waves to interfere constructively, we must have (the Bragg condition)

$$l\lambda = 2d \sin \phi, \tag{2.1}$$

where l is an integer. If the reflected wave is associated with a wave-vector $\mathbf{k}' = 2\pi\hat{\mathbf{u}}'/\lambda$, then it follows from Fig. 2.2 that the Bragg condition can also be expressed as a vector equation

$$\mathbf{d} \cdot (\mathbf{k} - \mathbf{k}') = 2\pi l, \tag{2.1a}$$

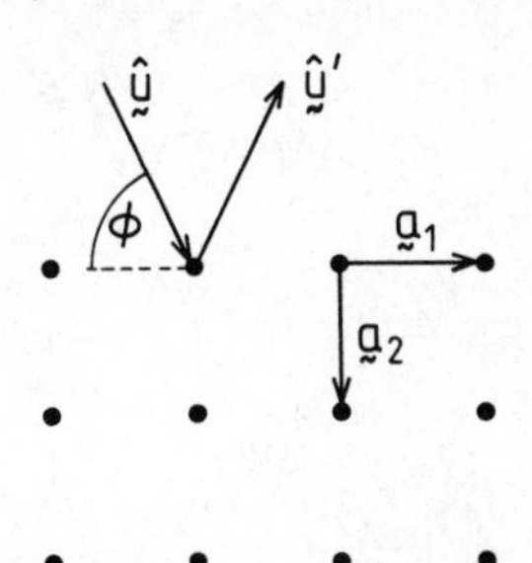

Fig. 2.1. A square lattice defined by a position vector $\mathbf{R} = n\mathbf{a}_1 + m\mathbf{a}_2$, showing incoming and outgoing plane waves parallel to unit vectors $\hat{\mathbf{u}}$, $\hat{\mathbf{u}}'$ and the angle ϕ between the lattice planes and $\hat{\mathbf{u}}$. The circles indicate positions of atoms in a crystalline lattice.

where $\mathbf{d}$ is now used as a position vector defining the lattice sites. Clearly, the constructive interference (a diffraction maximum) can only occur if the difference vector $\mathbf{G} = \mathbf{k} - \mathbf{k}'$ acquires certain values peculiar to the lattice in question, since from eqn (2.1a) we have the condition

$$\mathbf{G} = \frac{2\pi l}{\mathbf{d}}. \tag{2.2}$$

In general, the Bragg condition can be written for any vector $\mathbf{R}$ connecting two lattice sites in the form

$$\exp[\mathrm{i}(\mathbf{k} - \mathbf{k}') \cdot \mathbf{R}] = 1. \tag{2.3}$$

Since $|\mathbf{k}| = |\mathbf{k}'|$, the vector $\mathbf{G}$ must be perpendicular to the reflection plane (Fig. 2.3). Its length is inversely proportional to the length of the lattice vector $\mathbf{R}$ eqn (2.3). The vector $\mathbf{G}$ is defined in k-space, whereas $\mathbf{R}$ just defines positions of lattice points in real space, such as those shown in Fig. 2.1. Thus, to a set of points $\mathbf{R}$ in real space there is a unique set of

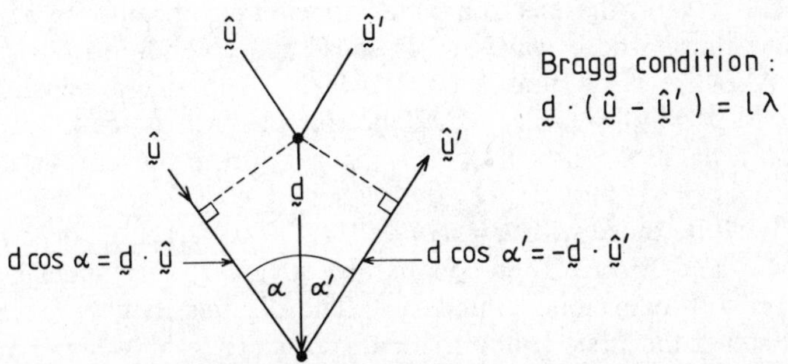

Fig. 2.2. A geometrical derivation of the Bragg condition in the structure of Fig. 2.1: λ is the wavelength of the diffracting beam; l is an integer.

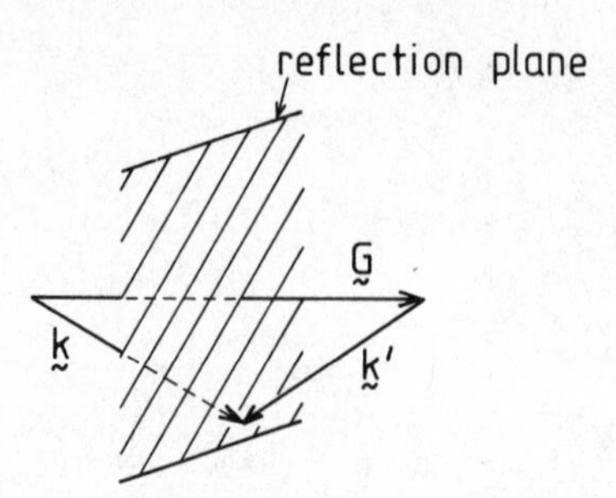

Fig. 2.3. The relations between the wave vectors $\mathbf{k}, \mathbf{k}'$ of the incoming and outgoing waves and the reciprocal lattice vector $\mathbf{G}$ (defined in the text), which is perpendicular to the reflection plane.

points (vectors $\mathbf{G}$) in k-space for which the Bragg condition is fulfilled. The vectors $\mathbf{G}$ are called reciprocal lattice vectors and the lattice of points in k-space defined by vectors $\mathbf{G}$ is called the reciprocal lattice. A diffraction peak therefore corresponds to a change in wave vector of the diffracted wave given by the reciprocal lattice vector $\mathbf{G}$. It is useful to remember the fact that the $\mathbf{G} \perp$ diffraction plane provides a convenient way to specify lattice planes. It is said that a plane normal to the reciprocal lattice vector $\mathbf{G} = h\mathbf{b}_1 + l\mathbf{b}_2 + m\mathbf{b}_3$ is characterized by Miller indices h, l, m and labelled (hlm). The Miller indices h, l, m are integers—with no common factor—inversely proportional to the intercepts (say x_1, y_1, and z_1) of the crystal plane along the crystal axes, i.e. $h:l:m = x_1^{-1}:y_1^{-1}:z_1^{-1}$. A few examples are shown in Fig. 2.4. A direction normal to a plane (hlm) is labelled $\langle hlm \rangle$.

Having described the electron interaction with the static lattice of ions as a diffraction process, we can now use the Bragg condition eqns (2.3) and (2.4) to find the electron states that will interfere most effectively. To minimize the amount of algebra needed here, we shall consider a one-dimensional crystal, i.e. an infinite array of atoms separated by a distance a. Again, the electron states affected significantly by the Bragg effect are those whose wave-vector satisfies the condition $\mathbf{G} = \mathbf{k} - \mathbf{k}' = 2\pi l/a$. Also, we know that $|\mathbf{k}'| = |\mathbf{k}|$ which in our one-dimensional case reduces to $|\mathbf{k}| = |\mathbf{k}'| = \mathbf{G}/2 = \pi l/a$. The free-electron states $\mathbf{k}$ available to us lie on a parabola $E_k = \hbar^2 k^2/(2m)$. Starting from the bottom at $k = 0$, we move upwards and inspect the values of k. The first pair of states affected will be those with $k = \pi/a$ and $k' = -\pi/a$. The next pair will be at $\pm 2\pi/a$, and so on. These points are obviously very special and are called the Brillouin zone boundaries. The distance between $-\pi/a$ and $+\pi a$ is called the first Brillouin zone, the distance between $-2\pi/a$ and $-\pi/a$ and between $2\pi/a$ and π/a is called the second Brillouin zone, and so on. These are shown in Fig. 2.5. We know that when the Bragg

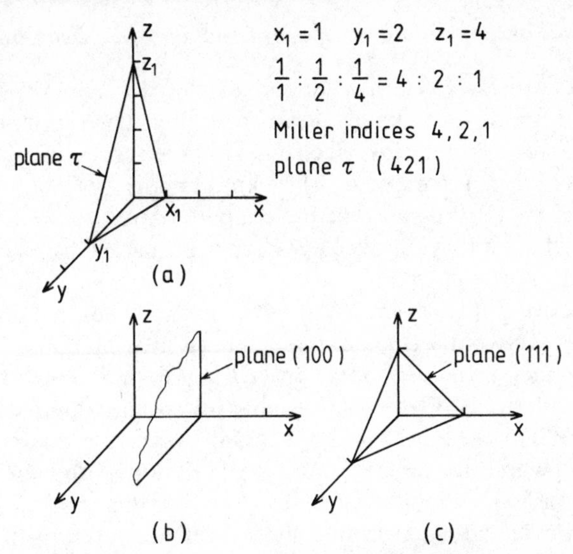

Fig. 2.4. (a) A plane intersecting real space (crystal) axes x, y, z at x_1, y_1, z_1. The Miller indices of this plane are 4, 2, and 1 and the label normally used for such a plane is (421). (100) and (111) planes are shown in (b) and (c), respectively.

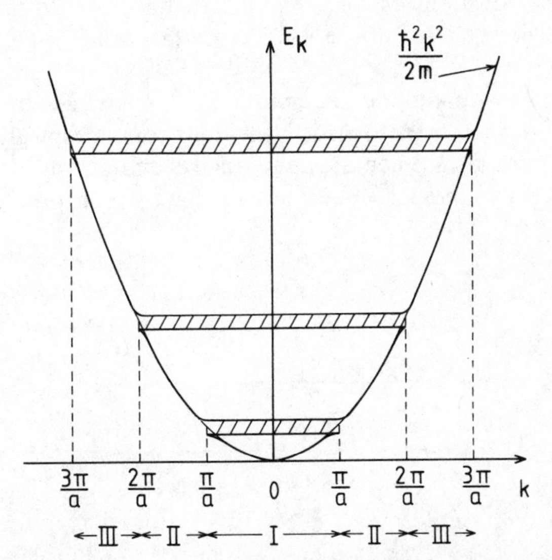

Fig. 2.5. The free-electron energy spectrum $E_k = \hbar^2k^2/(2m)$. Superimposed are the modifications at $k = \pm\pi/a$, $\pm 2\pi/a, \ldots$ introduced by Bragg reflections due to the crystal potential V_c. The curvature is changed in the vicinity of these points to join smoothly to the values of energy calculated at $\pm\pi/a$, $\pm 2\pi/a, \ldots$, which are $E_\pm = E_0 \pm |V_l|$, where E_0 is the corresponding free-electron energy and V_l is the lth Fourier component of V_c. I, II, and III are called the first, second, and third Brillouin zones, respectively. This picture is said to be an *extended zone representation* of the dispersion relation of E versus k.

condition is satisfied we no longer expect to get running waves $-\exp(ikx)$. Instead, we have total reflection corresponding to the interference effect, so that the electrons are now represented by standing waves of the form $\psi \sim \exp(\mathrm{i}\pi x/a) \pm \exp(-\mathrm{i}\pi x/a)$, i.e. $\psi_+ \sim \cos(\pi x/a)$ and $\psi_- \sim \sin(\pi x/a)$. We can plot the charge density $e\psi_+^*\psi_+$ and $e\psi_-^*\psi_-$ along the crystal axis x (Fig. 2.6). We can see that $|\psi_+|^2$ peaks at 0, $\pm a$, $\pm 2a$, . . . , i.e. at points where the atoms lie, whereas $|\psi_-|^2$ is zero there and peaks between the atoms. Without going into any details of atomic properties, we know that the atoms are represented by potentials that are attractive to electrons—the total (crystal) potential consisting of the overlapping potentials of the individual atoms is illustrated schematically in Fig. 2.6. When the solution is a running wave, the charge density is constant everywhere for all states and the variation of the strength of the attractive force exerted upon the electrons by this crystal potential is ignored. However, an electron in state ψ_+ will see more of the attractive crystal potential than that in state ψ_-, which will in fact see less than it would have seen had it been in a free-electron (non-interacting) state $\psi \sim \exp(\mathrm{i}\pi x/a)$, i.e. $|\psi|^2 = \text{constant}$. This means that we must expect state ψ_+ to lie lower in energy relative to the free-electron value, and state ψ_- higher, i.e. above the free-electron value of energy $\hbar^2(\pi/a)^2/(2m)$.

We can now estimate the magnitude of the change in the electron energy by returning to the Schrödinger equation of eqn (1.1) (which in our one-dimensional case contains only the second derivative with respect to x). Adding the crystal potential, we have, therefore, at the zone

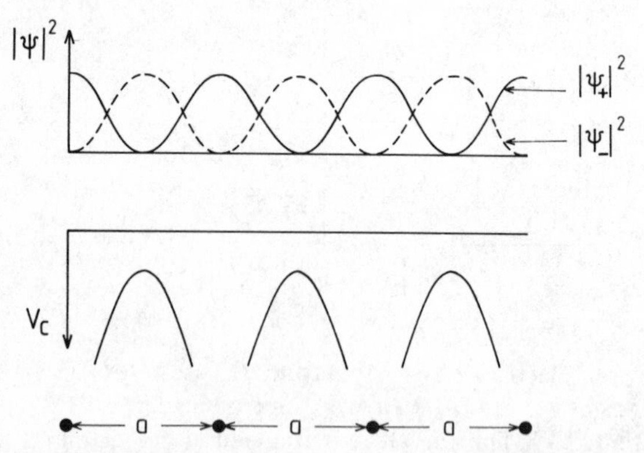

Fig. 2.6. A schematic representation of charge distribution associated with the lower (ψ_+) and higher (ψ_-) energy solutions of the Schrödinger equation at the Brillouin zone boundaries. V_c is the crystal potential associated with the positive ions (solid circles) separated by distance a. Note that $|\psi_+|^2$ peaks at the atoms, whereas $|\psi_-|^2$ peaks between them.

boundary,

$$\left\{-\frac{\hbar^2}{2m}\frac{d^2}{dx^2}+V_c\right\}\psi_\pm = E_\pm\psi_\pm, \tag{2.4}$$

where $\psi_+ = \text{const.} \times \cos(\pi x/a)$. Since the potential must be a periodic function of x, $V_c(x) = V_c(x+a) = V_c(x+2a), \ldots$, it must be possible to express $V_c(x)$ as a Fourier series,

$$V_c(x) = \sum_{l=-\infty}^{+\infty} V_l \exp(i2\pi lx/a) = \sum_G V_G \exp(iGx). \tag{2.5}$$

We can recognize in $2\pi l/a$ a familiar quantity, since it is the reciprocal lattice vector G for our one-dimensional lattice. In order to be able to find $E_\pm$ from eqn (2.4), we must ensure that our wave functions are properly normalized. We shall choose the normalization constant from the usual condition

$$\int_0^L \psi^*(x)\psi(x)\,dx = 1, \tag{2.6}$$

where L is the length of a unit box. L can be chosen arbitrarily large but remains finite, so that all mathematics can be nicely worked out. We obtain for ψ_+, E_+ (for the orthonormality of ψ_+, see eqns (2.11–2.13),

$$\left\{-\frac{\hbar^2}{2m}\frac{d^2}{dx^2}+\sum_{l=-\infty}^{+\infty} V_l \exp\left(i2\pi l\frac{x}{a}\right) - E_+\right\}\sqrt{\frac{2}{L}}\cos\left(\pi\frac{x}{a}\right) = 0. \tag{2.7}$$

We get the expectation value of E_+ by multiplying this equation from the left by $\sqrt{2/L}\cos(\pi x/a)$ and integrating over x from 0 to L. The first and the last terms are the ones we have determined before. To evaluate the kinetic energy term, we twice differentiate $\cos(\pi x/a)$ and integrate:

$$-\frac{\hbar^2}{2m}\frac{2}{L}\int_0^L \cos\left(\frac{\pi}{a}x\right)\frac{d^2}{dx^2}\cos\left(\frac{\pi}{a}x\right)dx = \frac{\hbar^2(\pi/a)^2}{2m}. \tag{2.8}$$

Using the normalization of ψ_+, we get for the last term

$$\frac{2}{L}\int_0^L \cos\left(\frac{\pi}{a}x\right)E_+\cos\left(\frac{\pi}{a}x\right)dx = E_+\frac{2}{L}\int_0^L \cos^2\left(\frac{\pi}{a}x\right)dx = E_+. \tag{2.9}$$

Finally, we have to evaluate the potential term. We shall write

$$\cos(\pi x/a) = [\exp(i\pi x/a) + \exp(-i\pi x/a)]/2,$$

so that this term becomes

$$\frac{1}{2L}\sum_{l=-\infty}^{+\infty}\int_0^L dx[\exp(-i\pi x/a) + \exp(i\pi x/a)]V_l \exp(i2\pi lx/a) \times [\exp(i\pi x/a) + \exp(-i\pi x/a)]. \tag{2.10}$$

We shall prove that

$$I = \frac{1}{L}\int_0^L \exp(-\mathrm{i}kx)\exp(\mathrm{i}k'x)\,\mathrm{d}x = \delta_{k,k'}. \tag{2.11}$$

If we recall that k must satisfy the periodic boundary condition, we can write $k = 2\pi m/L$ and $k' = 2\pi m'/L$ where m, m' are integers. Substituting into eqn (2.11) we obtain

$$I = \frac{1}{L}\int_0^L \exp\left[\mathrm{i}\left(\frac{2\pi m'}{L} - \frac{2\pi m}{L}\right)x\right]\mathrm{d}x = \frac{\exp[\mathrm{i}2\pi(m'-m)] - 1}{\mathrm{i}2\pi(m'-m)}. \tag{2.12}$$

If $k' = k$, then we obtain

$$\lim I_{m'\to m} = 1. \tag{2.13}$$

If $k' \neq k$ then $m' \neq m$, $\exp[\mathrm{i}2\pi(m'-m)] = 1$ and $I = 0$.

We can now use this result to eliminate those terms in eqn (2.10) for which argument of the complex exponential differs from zero. We find that finite contributions occur only for $l = \pm 1$. We shall ignore the term with $l = 0$ (which is also finite). V_0 is the mean value of the crystal potential. This term is the same for all states and leads to a constant shift of all energies; it is like shifting the whole of the parabola $\hbar^2k^2/(2m)$ in Fig. 2.5 up or down by V_0. Since the potential is real and symmetric, $V_{-1} = V_1$ in our case, and eqn (2.10) gives $V_1\ (L + L + 0 + 0)/(2L) = V_1$ so that we obtain for the energy

$$E_+ = E_0 + V_1, \tag{2.14}$$

where E_0 is the free-electron energy at $k = \pm\pi/a$. Thus if V_c is attractive to electrons, $V_1 = -|V_1|$ and the energy of state ψ_+ is lowered by $|V_1|$. We could now proceed along the same lines to show that the other solution, ψ_- at the Brillouin zone boundary ($k = \pm\pi/a$), is $E_- = E_0 + |V_1|$, so that this state is pushed up by the same amount. We could also extend our calculation to evaluate E_+ and E_- at the second, third, etc., Brillouin zone boundaries (i.e. at $k = \pm 2\pi/a$, $\pm 3\pi/a$, etc.). The result is again that there are two solutions, ψ_+ and ψ_-, one whose energy is lowered relative to the free-electron value, and the other whose energy goes up by $|V_2|$, $|V_3|$, ... Finally, we must ask what happens at $k \neq \pm\pi l/a$. We have assumed from the very beginning that the effect of the crystal potential is weak, so that V_1, V_2, ... are much smaller than the free-electron energy $E = E_k$ at the corresponding k. Far away from the zone boundary, the Bragg reflections are negligible and the wave function is just a running wave as in free-electron theory. Closer to the boundary, the reflections are not negligible and the wave function must also include the reflected wave, e.g., $\psi_k \sim a_1\exp(\mathrm{i}kx) + a_2\exp(\mathrm{i}k'x)$ where a_1, a_2 must be determined by a more detailed quantum-mechanical calculation. For states where a_1 and a_2 have the same sign the energy will be lowered, whereas for states where the signs are opposite the energy would go up, and each would join smoothly with the corresponding values (E_+ or E_-) of energy obtained in our calcuation for k at the zone boundary. This is illustrated in Fig. 2.5. This figure shows that allowed solutions of the Schrödinger equation form bands of energies separated by gaps in the energy spectrum where there are no solutions. The magnitude of these forbidden gaps is given by the Fourier components V_l of the crystal potential associated with the non-zero reciprocal lattice vectors. Other ways of arriving at the energy band diagram are outlined in an example in Appendices 2 and 3, where a specific case of an array of one-dimensional barriers (the Krönig–Penney problem) is presented.

The diagram in Fig. 2.5 is known as the extended zone representation. We can arrange the same set of results in a more compact way. (Fig. 2.7). Since the lengths of the first, second, third, etc., Brillouin zone are the same, we can fold all our results into the interval $-\pi/a$ to π/a, i.e. into the first zone. The energy bands remain, of course, exactly the same, since we have merely decided to redraw Fig. 2.5 by shifting higher states by a reciprocal lattice vector into the first zone. There are now many different energy levels at any given wave vector k in the first zone and we need another index to distinguish them from each other. We say that the lowest energy belongs to band 1, the next higher up to band 2, and so on. Hence, any solution can be unambiguously identified in terms of two 'quantum numbers', i.e., by the value of k in the first zone (the so-called reduced wave vector) and by the band index n. This way of presenting

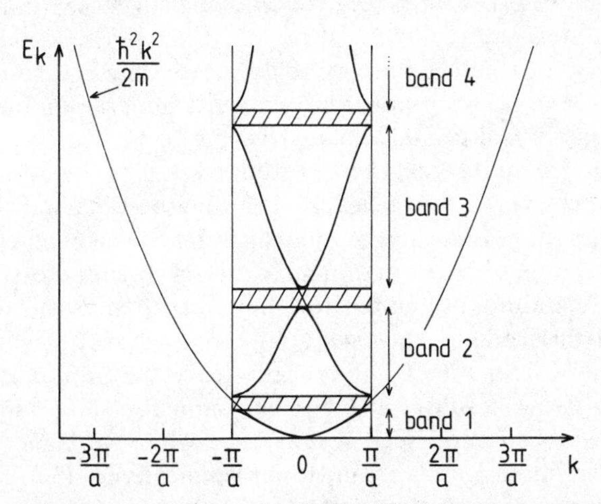

Fig. 2.7. The diagram of E versus k of Fig. 2.5 folded into the first Brillouin zone. The gaps are indicated by hatched zones and bands are numbered 1, 2, 3, and 4. This is the *reduced zone representation* of band structure.

our results is called the reduced zone representation. The picture of energy bands versus reduced wave vector is called the band structure.

We can now examine our results in more detail. For example, we can ask how many electrons can occupy one band. We know that because of the periodic boundary condition the volume per quantum state of a given allowed wave vector for our one-dimensional crystal is $2\pi/L$. The volume of the Brillouin zone is $2\pi/a$, so that there are L/a states in a band. But a is the nearest-neighbour distance and L is the length of the 'unit' cell so that L/a is the number of atoms in L. Each quantum state can accommodate two electrons, so that we find that each band can take up to two electrons from each atom. This is an important result. Since bands are separated by energy gaps, electrons cannot jump from one band to another without an external source of energy. If our solid consists of atoms that possess an even number of electrons say, four, which is the case in silicon, then the lowest two bands will be completely full and the remaining bands 3, 4, 5, etc., entirely empty. Hence, our theory predicts that silicon (indeed, any solid consisting of atoms with an even number of valence electrons) is an insulator at low temperatures. Similarly, a solid consisting of atoms each of which has an odd number of valence electrons (e.g. sodium) is predicted to behave as a metal (conductor), because in this case the uppermost band containing electrons is only half full. Since the energy levels belonging to adjacent allowed values of k are infinitesimally close to each other, an electron can easily jump from an occupied state to the nearest empty one. It is only when electrons can jump from occupied states to empty ones that they can, upon application of a directional external force (e.g. an electric field) flow from one end of the crystal to another.

Although in our calculation we have used only a one-dimensional model crystal, the above predictions are quite general and applicable to real three-dimensional crystals. Naturally, the band structure for a given solid, i.e. an accurate diagram of the dispersion relations showing electron energies versus wave vector, can only be obtained if a detailed computer calculation is carried out in which the true symmetry properties of the crystal and the strength and form of the microscopic crystal potential are carefully accounted for. Such calculations lie well outside the scope of this course. However, our simple model provides us with concepts that will enable us to benefit from the results of full-scale calculations. These results for key semiconductors are summarized in Chapter 5. We shall need such results later when we come to consider semiconductor microstructures and their applications. However, before we start looking at 'real' semiconductor band structures we must return to our one-dimensional model and generate, in the next chapter, a few new concepts that will complete our classification of solids.

Summary

In the nearly-free-electron model, electrons are allowed to interact weakly with the lattice of positive ions. This interaction is governed by the Bragg law and leads to diffraction maxima (standing waves) when the wave vectors of the incoming and outgoing waves differ by a reciprocal lattice vector **G**. The resulting electronic structure differs from the free-electron one at the Brillouin zone boundary, where zones of forbidden energies (gaps) separate bands of allowed energies. Each band can accommodate two electrons from each atom, so that at low temperatures crystals consisting of atoms with an even number of electrons are predicted to behave as insulators and those with an odd number to behave as metals (conductors). The magnitude of the forbidden gap depends on the magnitude of the Fourier components V_G of the crystal potential. In particular, the magnitude of the gaps of a one-dimensional crystal (E_g) at the Brillouin zone boundary $k = G/2$ is $2\,|V_G|$. For the model to remain valid E_g must be small compared to the free-electron energy at the Brillouin zone boundary.

Problems

2.1. Find the cosine of the angle between planes $(u_1u_2u_3)$, $(v_1v_2v_3)$ in a cubic lattice.

2.2. Show that the distance between adjacent planes (hkl) in a cubic lattice is $d = a\,(h^2 + k^2 + l^2)^{-1/2}$, where a is the lattice constant. Calculate the magnitude of d for the (100) planes of silicon.

2.3. Show that the expectation value of linear momentum vanishes for a particle at the band edge of a nearly-free-electron crystal.

2.4. A one-dimensional periodic potential has Fourier coefficients $V_n = V_1/n^2$. The width of the tenth nearly-free-electron gap is 0.032 eV. What is the magnitude of V_1?

2.5. Consider a simple cubic lattice in which the atomic separation is 2 Å. Determine the electron density from the condition that $2k_F = y$, where y^3 is the volume of the first Brillouin zone.

2.6. Consider the result of Problem 2.4 and determine at what range of wavelengths this crystal is transparent. What would the value of V_1 have to be for the crystal to be transparent to red light?

3

Localized representation of electron states in solids: tight-binding model

The nearly-free-electron model correctly accounts for the existence of simple metals such as sodium, potassium, or aluminium, and of insulators whose forbidden gap separating the occupied valence band from the empty conduction band is small compared to the free-electron energy at the Brillouin zone boundary (e.g. silicon, germanium, GaAs, and all other so-called semiconductors). There are, however, important materials for which this theory is not applicable. In these materials, the electron wave functions from the adjacent atoms in the lattice do not overlap very strongly and, in contrast with the assumptions of the nearly-free-electron model, retain much of their atomic character, i.e. they are tightly bound.

To develop a model for solids with tightly bound electronic wave functions, let us turn to a one-dimensional example. We must construct a new crystal wave function so that it would, at least in the limit of large separation between atoms, turn itself into a superposition of identical free-atom wave functions, localized at the lattice site and representing the atom sitting there. Consider an array of N atoms separated by distance a. Let us write the electronic wave function ψ for this one-dimensional solid in the form

$$\psi(x) = N^{-1/2}\sum_{j=1}^{N} c_j\phi(x - x_j), \tag{3.1}$$

where

$$\int_{-\infty}^{\infty} \phi^*(x)\phi(x)\,\mathrm{d}x = 1, \tag{3.2}$$

j labels atoms in the lattice, and $\phi(x - x_j)$ is a free-atom wave function at the jth site (ϕ satisfies the Schrödinger equation for a free atom with electron energy E_{at}). If the atoms were entirely independent, the coefficients c_j in the expansion for ψ in eqn (3.1) would be equal to unity. The condition in eqn (3.2) and the factor $N^{-1/2}$ ensure that both ψ and

$\phi(x - x_j)$ are normalized to unity, as usual in quantum physics. (If, for example, we chose to construct this lattice of hydrogen atoms, then ϕ would be the wave function of the hydrogen ground state 1s.)

We shall determine coefficients c_j from the condition that ψ must have the same general symmetry properties as the wave functions for periodic structures considered in the nearly-free-electron model. In particular, we shall require that

$$\psi(x + R) = \exp(\mathrm{i}kR)\ \psi(x), \tag{3.3}$$

where R is a multiple of the nearest-neighbour distance a (i.e. a translation vector). It is easily verified that if we take ψ to be a linear combination of planes waves, this condition is satisfied. When

$$\psi \equiv \psi_k = \text{const.} \exp(\mathrm{i}kx), \tag{3.4}$$

then we obtain (k is the wave vector)

$$\psi_k(x + R) = \text{const.} \exp(\mathrm{i}k(x + R)), \tag{3.5}$$

which can be written as

$$\psi_k(x + R) = \exp(\mathrm{i}kR)[\text{const.} \exp(\mathrm{i}kx)] = \exp(\mathrm{i}kR)\psi_k(x). \tag{3.6}$$

The condition in eqn (3.3) is indeed satisfied. It is possible to prove quite generally that any solution of the Schrödinger equation with a periodic potential must obey eqn (3.3). (This is the so-called Bloch Theorem.)

Let us now choose ψ in eqn (3.1) with $c_j = \exp(\mathrm{i}kx_j)$. We can verify that this wave function is also well behaved.

$$\begin{aligned}\psi \equiv \psi_k(x + R) &= \sum_j \exp(\mathrm{i}kx_j)\ \phi(x + R - x_j) \\ &= \sum_j \exp(\mathrm{i}kR) \exp[\mathrm{i}k(x_j - R)]\ \phi(x - (x_j - R)).\end{aligned} \tag{3.7}$$

Let us relabel lattice sites by defining $x_l = x_j - R$. Substituting into eqn (3.7), we obtain

$$\psi_k(x + R) = \exp(\mathrm{i}kR)\sum_l \exp(\mathrm{i}kx_l)\ \phi(x - x_l) = \exp(\mathrm{i}kR)\psi_k(x). \tag{3.8}$$

The Schrödinger equation for the one-dimensional solid is, with ψ of eqn (3.7),

$$\left\{-\frac{\hbar^2}{2m}\frac{\mathrm{d}^2}{\mathrm{d}x^2} + V_c - E_k\right\}\frac{1}{\sqrt{N}}\sum_{j=1}^{N} \exp(\mathrm{i}kx_j)\ \phi(x - x_j) = 0. \tag{3.9}$$

In order to obtain the energy (E_k), we proceed as before: we must multipy eqn (3.9) from the left by the complex conjugate of the wave function ψ and integrate over x:

$$\frac{1}{N}\sum_{n=1}^{N}\sum_{j=1}^{N}\int_{-\infty}^{\infty}\Big[\exp(-ikx_n)\,\phi^*(x-x_n)\Big[-\frac{\hbar^2}{2m}\frac{d^2}{dx^2}+V_c-E_k\Big]$$

$$\times\exp(ikx_j)\,\phi(x-x_j)\Big]\,dx=0. \quad (3.10)$$

The crystal potential V_c is a superposition of atomic potentials,

$$V_c=\sum_{l=1}^{N} v_{\text{atom}}(x-x_l). \quad (3.11)$$

We shall assume that only the states from adjacent atoms interact:

$$\int_{-\infty}^{\infty}\phi^*(x-x_n)\Big\{-\frac{\hbar^2}{2m}\frac{d^2}{dx^2}+V_c\Big\}\phi(x-x_j)\,dx=\begin{cases}-\beta,\ n=j\pm1,\\ E_{\text{at}},\ n=j,\\ =0,\ n\neq j\neq j\pm1.\end{cases} \quad (3.12)$$

Finally, we shall assume that explicit overlap integrals can be neglected, i.e.

$$\int_{-\infty}^{\infty}\phi^*(x-x_n)\,\phi(x-x_l)\,dx=\begin{cases}1,\ \ldots,\ n=l,\\ 0,\quad n\neq l.\end{cases} \quad (3.13)$$

When $n=l$, we recover the normalization condition for the free-atom wave function ϕ.

With these simplifications, eqn (3.10) reduces to

$$E_k=E_{\text{at}}-\beta(\exp(-ika)+\exp(ika)), \quad (3.14)$$

or, in a more usual form, to

$$E_k=E_{\text{at}}-2\beta\cos(ka). \quad (3.15)$$

The factor N^{-1} disappeared because there are $2N$ identical nearest-neighbour integrals, two for each of the N lattice sites. There are also N integrals yielding E_{at} and E_k (one for each site). We obtained the simple form of eqn (3.15) because we neglected all integrals in eqn (3.10) representing interactions between more distant (i.e. other than nearest-neighbour) atoms. This is well in keeping with our original intentions here, since we want to study crystals in which valence electron wave functions are tightly bound to the atom they originated from, and interactions between atoms are weak.

The expression for energy E_k in eqn (3.15) depends on two parameters, which represent the strength of the nearest-neighbour interaction peculiar to the atomic valence state (β), and the separation of the lattice sites (atoms) a. Since $\cos(ka)$ varies from -1 to 1, all solutions derived from the free-atom valence state at E_{at} lie in a band of energies of width 4β. Of course, it is clear from eqn (3.12) that β also depends on a and is expected to increase as a decreases.

A band can accommodate two electrons from each atom in the lattice. This is now intuitively even more apparent, since in the limit $a \to \infty$ (independent atoms), the atomic state $\phi(x - x_j)$ of eqn (3.1) can take two electrons and is available on each site j. There are, therefore, N allowed values of k (or N states) in each band, as follows also from the quantization of the wave vector we became accustomed to in the nearly-free-electron model.

We could generalize our theory by picking another atomic state, with a different atomic energy E'_{at} and repeating the calculation, with formally identical results. The new wave function ψ' would give rise to a new value of the nearest-neighbour interaction parameter, β', and the band width of our second, higher-lying band, would be $4\beta'$. For example, if we took ϕ to be the ground state of hydrogen, we could then take ϕ' to be the first excited (2s) state, and with suitable β and β' we would obtain the lowest two bands of the band structure for a one-dimensional hydrogen crystal.

Our general results of eqn (3.15) are summarized in Fig. 3.1, where we can also see a new interesting feature. Suppose that we decrease the nearest neighbour separation a so much that the wave function overlap

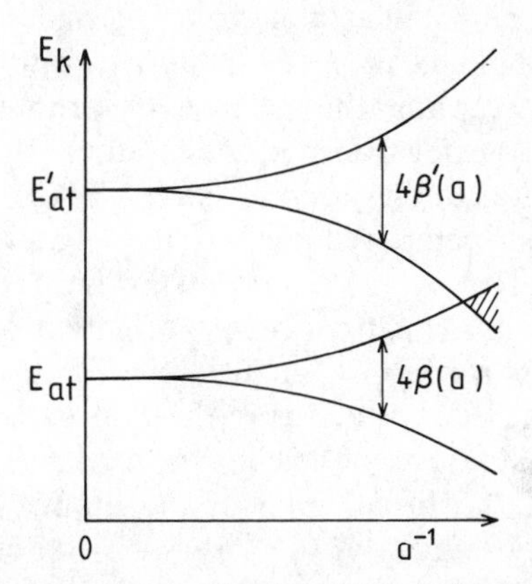

Fig. 3.1. The energy of an electron in the tight-binding model described in eqn (3.15): a is the separation of the atoms (lattice constant). E_{at} and E'_{at} are the free-atom energies obtained for the lowest two (atomic) states corresponding to the limit $a \to \infty$. For smaller interatomic distances, the adjacent atomic wave functions overlap and the energy levels are broadened into bands of widths 4β and $4\beta'$, respectively.

between nearest neighbours becomes large and the band widths 4β, $4\beta'$ increase. Figure 3.1 implies that at a certain critical separation, adjacent bands of allowed energies overlap. This destroys the forbidden energy gap between the two bands. Let us assume that we have a crystal with two electrons per atom. Normally we expect such a crystal to be an insulator at low temperatures. However, when the valence and conduction bands overlap, the uppermost electrons can easily jump into a nearby empty state. The model therefore predicts that under certain circumstances it is conceivable to have metals consisting of atoms with an even number of valence electrons. Such metals do in fact exist (e.g. calcium).

The main importance of the tight-binding model lies in its power to provide a way of modelling periodic systems with narrow bands, i.e. with band width 4β being small compared to the energy difference $E_{\text{at}} - E'_{\text{at}}$ between free-atom states ϕ and ϕ'. In such a case the 'band gap' is not very different from the separation of free-atom energy levels. Hence the picture of the electronic structure we arrived at in this model is precisely the opposite of what we obtained from the nearly-free-electron model.

Let us now identify the class of solids that are best described by the tight-binding model. For the valence electron in a solid to remain tightly bound (atomic-like), it must either accomplish a complete jump to a nearest-neighbour atomic state or stay in its free-atom state. Inert atoms with complete outer shells (from the eighth column of the Periodic Table), such as neon, form solids of the latter type. The former type fits a situation where the neighbour atom has just one slot available in the outer shell. For example, in NaCl, chlorine can complete its outer (valence) shell by accepting the valence electron of its neighbouring sodium. The charge transfer from the cation atom (Na) to the anion atom (Cl) is nearly complete. In the jump 3s(Na)$\rightarrow$3p(Cl), the Na–Cl pair can lower its total energy compared to the sum of independent free-atom energies by ~7 eV. The energy diagram showing the uppermost atomic levels of Na and Cl participating in the formation of the NaCl bond is shown in Fig. 3.2. The energy gained in the jump is a measure of the 'ionicity' of the bond between atoms in the lattice.

If we now form a lattice of NaCl, we expect, in the spirit of the tight-binding model, the broadening of the atomic 3p(Cl) state into a band to be small. Consequently, the forbidden gap of NaCl should be of order 7 eV. This is indeed the case.

Crystals with a similar strong ionic bond can be formed by other atoms from the first and seventh columns of the Perioidc Table (the so-called I–VII compounds). When we come to consider II–VI and III–V compound crystals, such as, for example, CdTe and GaAs, respectively, the difference between the corresponding atomic energy levels of the

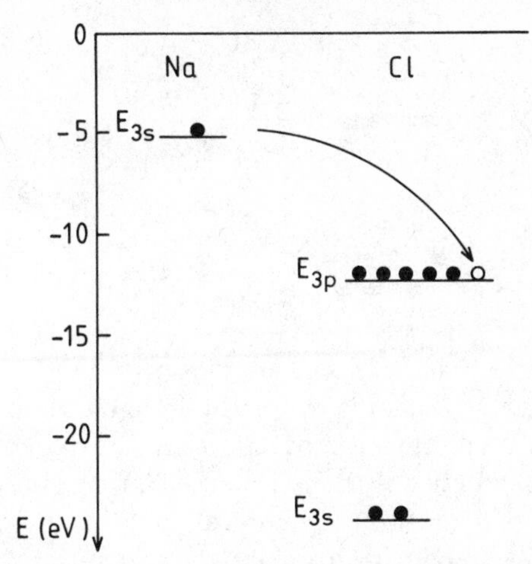

Fig. 3.2. The energy levels of the valence electrons of sodium and chlorine atoms. When a crystal of NaCl is formed, the 3s electron of sodium is transferred to the 3p state of chlorine and completes its valence shell.

cation and anion atoms becomes smaller, and the charge transfer is incomplete. In group IV crystals of silicon or germanium, this energy difference is reduced to zero, since all atoms in the lattice are the same. The transfer of charge from atom to atom does not occur. Instead, the valence electrons are 'shared' in what is called a purely 'covalent' bond. The charge is transferred from atoms towards the midpoints of the shortest path connecting nearest neighbours. Hence, it is not very surprising that wave functions in covalent semiconductors like silicon are delocalized and can be described by the nearly-free-electron model.

In our account of the electronic band structure, the emphasis has been on those parts of the energy spectrum where the Bragg maxima and minima occur, namely at the Brillouin zone boundaries. In a non-metallic crystal, these are the points where the states occupied by the uppermost valence electrons lie, separated from the nearest empty states by a forbidden energy gap. We shall see that most of the processes of interest (e.g. conduction, absorption of light) occur in this range of the electron energy spectrum. In order to simplify the description of such processes, we shall use a parametrized form for the curve E_k, valid at the band minima or maxima (the band 'edges'). Let us expand the function E_k say, at a minimum that may occur at a point $k = k_0$, as a power series:

$$E_k|_{k_0} = E_{k_0} + A(k - k_0) + B(k - k_0)^2 + \cdots. \tag{3.16}$$

Since close enough to k_0 we expect $E_k = E_{-k}$, we set $A = 0$, so that near k_0

the functional form of E_k is given by $B(k-k_0)^2$. Let us choose B so that

$$E_k = \text{const.} + \frac{\hbar^2(k-k_0)^2}{2m^*},$$

where

$$m^* = \hbar^2\left(\frac{\partial^2 E_k}{\partial k^2}\right)^{-1} = \frac{\hbar^2}{2B} \tag{3.17}$$

is called the effective mass. Hence, any deviation of the dispersion curve E_k at band minima or maxima from the free-electron functional form $\hbar^2k^2/(2m)$ can be accounted for by choosing the relevant value of m^*. The particle at a band edge is then viewed as a free quasi-particle with an 'effective' mass m^*, and all dynamical relations available for free particles can still be made use of simply by replacing the free electron mass m with m^*. The value of m^* can be obtained experimentally and from calculations. The magnitude of m^* represents the strength of the effect of the crystal potential.

Summary

The tight-binding model of electronic structure is based on the assumption that the electron wave function in the crystal remains localized at the lattice sites and the wave function overlap between sites is small. This model predicts that the electron energy E_k for a one-dimensional crystal depends on the degree of nearest-neighbour interaction β and on the nearest-neighbour distance a as $E_k = E_{at} - 2\beta\cos(ka)$, where k is the wave vector and E_{at} is the free-atom energy level the valence electron would possess in the limit $a \to \infty$. The model offers the possibility of completing the classification of solids.

We introduced a simplified representation of the dispersion relations E_k near the band edge by defining the effective mass $m^* = \hbar^2(\partial^2E_k/\partial k^2)^{-1}$. In particular, very close to a band minimum at $k=0$, the energy is simply $E_k = \hbar^2k^2/(2m^*)$.

Problems

3.1. Find the expression for the group velocity of an electron in a narrow tight-binding band and compare the result with that for a free-electron metal and a nearly-free-electron insulator.

3.2. Find the effective mass of an electron in a tight-binding band as a function of the atomic separation a.

3.3. Extend the result in eqn (3.15) to the case of a three-dimensional simple cubic lattice.

3.4. Predict general features of the electronic structure of Si, GaP and ZnS crystals from the free-atom valence electron (s and p) energies given below.

$-E$(eV)	Zn	Ga	Si	P	S
s	8.40	11.37	13.55	17.10	20.80
p	3.38	4.90	6.52	8.33	10.27

4

Basic observable properties of semiconductor materials

4.1. Thermal excitations. Electron density in the conduction band. Law of mass action

We have shown that at zero temperature ($T = 0$) the conduction band of a material like silicon is entirely empty and the valence band is full. If we raise the temperature, atoms begin to vibrate vigorously and this supplies energy to valence electrons. There is then a finite probability that some of them gain enough energy to overcome the fundamental energy gap E_g separating the top of the valence band and the bottom of the conduction band. This probability is given by the Fermi–Dirac distribution function, f_{FD}, (Fig. 4.1):

$$f_{\mathrm{FD}}(E) = [\exp\{(E - E_F)/k_B T\} + 1]^{-1} \tag{4.1}$$

(k_B is the Boltzmann constant), which is the probability of finding an electron at energy E above the top of the electron reservoir located at E_F. We know that in silicon there are no electrons at the free-electron Fermi energy E_F. We showed in Chapter 2 that the free-electron states from the range of energies E_F to $E_F + \frac{1}{2}E_g$ were pushed by the crystal potential up to the conduction band edge, and those from the range E_F to $E_F - \frac{1}{2}E_g$ were pushed down into the valence band, so that there are no allowed states available to electrons in the forbidden gap (Fig. 4.2). We can, however, still think of E_F as a point in the gap and view an excitation from the top of the valence band to the conduction band as a process of taking an electron from E_F to $E_F + \frac{1}{2}E_g$, and taking the hole, created at the top of the free electron reservoir, from E_F down by $\frac{1}{2}E_g$ to the valence band. Usually, $E - E_F \gg k_B T$ ($E_g(\mathrm{Si}) \approx 1\,\mathrm{eV}$; at $T = 300\mathrm{K}$, $k_B T \approx 25\,\mathrm{meV}$) and f_{FD} reduces to the classical Boltzmann form

$$f_B(E) = \exp[-(E - E_F)/k_B T]. \tag{4.2}$$

Near the band edge, the functional form of E_k can be approximated by a free-electron parabola, with m replaced by the corresponding effective

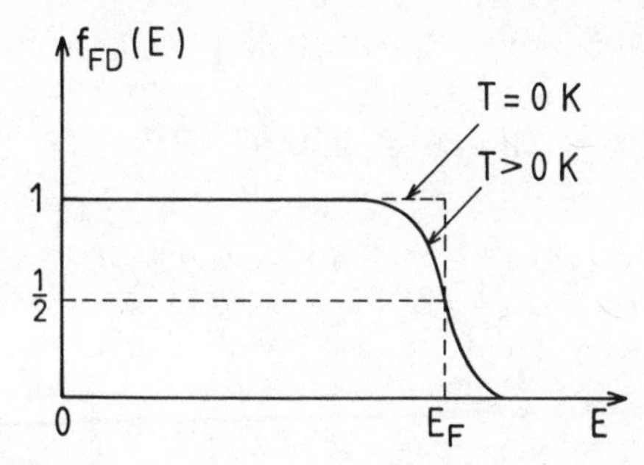

Fig. 4.1. The probability $f_{FD}(E)$ that a state at energy E is occupied at temperature T (eqn (4.1)). E_F is Fermi energy defined in Chapter 1.

mass introduced in Chapter 3. Hence we have $E_k = \hbar^2k^2/(2m^*)$ at the edge, so that the density of states in the conduction band (the number of electron states per unit energy interval per unit volume) is

$$\rho^e(E) = \frac{1}{2\pi^2}\left(\frac{2m_e^*}{\hbar^2}\right)^{3/2}(E - E_g)^{1/2}. \tag{4.3}$$

The energy is measured from the top of the valence band. The number of occupied states in the conduction band in an interval dE is d$n(E) = f_B^e(E)\rho^e(E)\,\mathrm{d}E$ and the total electron concentration n in the band is

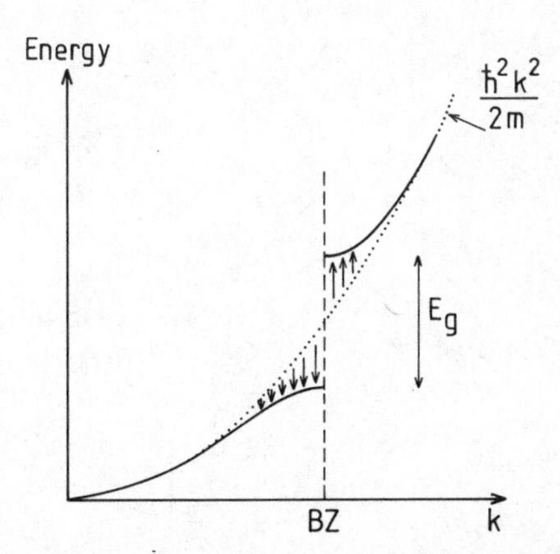

Fig. 4.2. The effect of a one-dimensional crystal potential upon free-electron states lying in the range of energies where the forbidden gap is situated. BZ is the Brillouin zone boundary. In the limit of small gap E_g relative to the free-electron Fermi energy, $k_F \simeq k_{BZ}$. The uppermost occupied free-electron states are suppressed by the effect of the crystal potential into the valence band and the empty states in the range of energies E_F to $E_F + \frac{1}{2}E_g$ are pushed into the conduction band.

obtained by integrating over all conduction band energies, i.e.

$$n = \int_{E_g}^{\infty} f_B^e(E)\rho^e(E)\,\mathrm{d}E. \tag{4.4}$$

Although m_e^* is valid only at band edge, f_B decreases rapidly with E and only the region close to E_g contributes. $n(E)$ is shown in Fig. 4.3 We can substitute for f_B^e and $\rho^e(E)$ to obtain

$$n = \frac{1}{2\pi^2}\left(\frac{2m_e^*}{\hbar^2}\right)^{3/2} \int_{E_g}^{\infty} \exp[-(E - E_F)/k_BT](E - E_g)^{1/2}\,\mathrm{d}E. \tag{4.5}$$

This reduces to a standard integral $\int_0^\infty y^{1/2}\mathrm{e}^{-y}\,\mathrm{d}y = \pi/2$, so that

$$n = 2\left(\frac{m_e^* k_B T}{2\pi\hbar^2}\right)^{3/2} \exp(E_F/k_BT)\exp(-E_g/k_BT). \tag{4.6}$$

Following the same procedure, we obtain for the hole concentration in the valence band at temperature T,

$$p = 2\left(\frac{m_h^* k_B T}{2\pi\hbar^2}\right)^{3/2} \exp(-E_F/k_BT). \tag{4.7}$$

We must remember that m^* must be found separately for each band extremum.

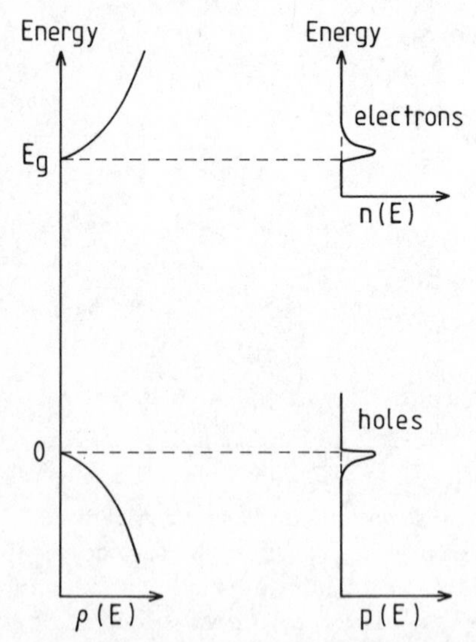

Fig. 4.3. (Left) The density of states near the conduction and valence band edges. (Right) The electron and hole densities n and p in the conduction and valence bands, respectively, at finite temperatures.

Since by exciting an electron we create a hole, we must have

$$f^h = 1 - f^e; \qquad n = p. \tag{4.8}$$

By equating eqns (4.7) and (4.6) we can recover E_F as a function of temperature:

$$E_F = \tfrac{1}{2}E_g + \tfrac{3}{4}k_B T \ln(m_h^*/m_e^*). \tag{4.9}$$

As $T \to 0$, we obtain $E_F = \frac{1}{2}E_g$, which is consistent with our starting point (Fig. 4.1). The value of $n = p = n_i$ is the so called intrinsic carrier concentration. If we take the product np we find

$$n_i^2 = np = 4\left(\frac{k_B T}{2\pi\hbar^2}\right)^3 (m_e^* m_h^*)^{3/2} \exp(-E_g/k_B T). \tag{4.10}$$

This 'law of mass action' shows that n_i is independent of the valence electron density in the material, since E_F disappears from the formula. Therefore, the number of free carriers we can create in a given material at temperature T depends only on the band structure parameters m_e^*, m_h^* and E_g. Any changes in n are exactly compensated for by changes in p. Note that if E_g is sufficiently small, the number of free carriers will initially increase with increasing temperature. Since the current density $j \sim n$ (see Section 4.7), j will also increase, until at higher temperatures the collisions with lattice waves excited by high T dominate and reduce j. This behaviour of j versus T does not occur in good conductors (in metals $E_g = 0$), where j decreases with increasing T. In insulators with large E_g, n is negligibly small except at very high T. Hence, crystals with small E_g ($\approx$0–2 eV) are called semiconductors.

4.2. Lattice vibrations. Phonons

The models of electrons states described in preceding chapters have assumed a static lattice. In fact, even at very low temperatures, atoms vibrate around the static lattice sites we have assigned them. Of course, atoms are more than 2000 times heavier than electrons and their motion is therefore so slow that electrons can follow effectively instantaneously. Also, the amplitudes of these vibrational displacements are negligibly small compared to the distances between atoms, because atoms are held in position by strong elastic forces that we can imagine as springs connecting the nearest neighbours in the lattice. We can specify the spring constants and use Hooke's law (displacement $\propto$ applied force) to evaluate the energy versus frequency of the vibrational motion, using equations of motion familiar from elementary classical mechanics. The vibrational waves are characterized by wave vectors $\mathbf{k}$ that must satisfy

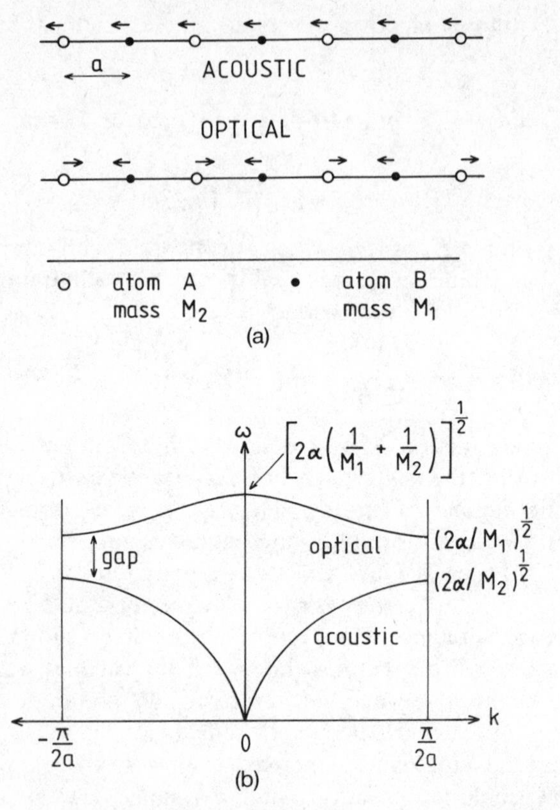

Fig. 4.4. Vibrating one-dimensional lattice consisting of two different atoms A and B. Acoustic and optical modes of vibrations are shown in (a). (b) The dispersion relations for vibrational waves (energy $E = \hbar\omega$ versus wave vector) obtained in Problem 4.3. Note that this phonon spectrum has features similar to the energy spectrum for electrons in such a lattice.

the same periodic boundary conditions as the electron waves. Since $E \sim m^{-1}$, the energy of a typical vibrational wave is of order $E_\mathbf{k}$(electron) $\times\, m$(electron)$/m$(proton), i.e. about 10 meV. A vibrational wave of a given $\mathbf{k}$, $(E_\mathbf{k} = \hbar\omega_\mathbf{k})$ is treated as a particle and called a phonon. A typical dispersion relation is shown in Fig. 4.4.

4.3. Optical absorption and emission. Direct and indirect transitions

Let us shine light of wavelength λ and energy $E = \hbar\omega = hc/\lambda$ onto a crystal with a forbidden gap E_g at zero temperature (i.e. no thermally excited electrons and holes). We can place a detector behind the sample and ask what we should observe. Since energy must be conserved, if the uppermost valence electron receives in the collision with a photon (i.e. a quantum of light of energy hc/λ) less than energy E_g, it will not be able to reach the nearest empty state. Our detector will register all incoming light that we shone on the sample. We say that the sample is transparent at this wavelength. When $hc/\lambda \geqslant E_g$, the electrons in the valence band absorbing the photon energy receive enough energy to move them into the conduction band. Our detector will register that no light is going through the sample. If we apply a weak electric field across the sample, we can measure the current of the photoexcited electrons. Of course, we must continue to shine light, since the excited electrons, having reached the conduction band, will only stay there on average for a short interval of time (the electron lifetime) and this depletes the supply of free carriers. After this time the conduction electron recombines with one of the holes in the valence band, i.e. it emits light (a photon) of energy E_g and jumps down into the valence band. Apart from the fact that energy must be conserved in the photon absorption as well as in emission processes, we also have to conserve linear momentum. The momentum of an electron near the valence band edge is of order $p = \hbar k_{\mathrm{BZ}}$ where $k_{\mathrm{BZ}} = \cong \pi/a \sim 10^8\ \mathrm{cm}^{-1}$. The photon wave vector for light in the visible range is $k_{\mathrm{ph}} \sim 10^5\ \mathrm{cm}^{-1}$ so that for light in the visible and lower wavelength range we can assume that the photon momentum is negligibly small. This means that the momentum conservation rule for the photon collision with a valence (or conduction) electron prescribes that the electron momentum in the initial state must be the same as in the final state, i.e. it must remain uneffected by the collision. Consequently, an optical (or radiative) transition corresponds to a vertical jump in the band structure diagram of E_k versus k, the so-called direct transition.

The band structures of technologically important materials do differ in detail from the simple diagrams presented in Chapter 2. This is because a larger number of the Fourier components of the crystal potential must be taken into account in the calculation of E_k versus k. It turns out that in some cases (e.g. silicon) the lowest point of the conduction band curve does not occur at the same point of the Brillouin zone as the top of the valence band (see Chapter 5). In that case, the momentum conservation cannot be satisfied, and the jump across the band gap from the top of the valence band can only be achieved if the absorbing or emitting electron collides with a vibrational lattice wave (phonon) of suitable momentum k_{ph} such that $\mathbf{k}_v + \mathbf{k}_{\mathrm{ph}} = \mathbf{k}_c$, where k_v and k_c indicate the position of the valence band maximum and the conduction band minimum in the first Brillouin zone, respectively (Fig. 4.5). This phonon-assisted or indirect transition is, of course, less likely than a direct one because it depends on the probability that our electron collides with a suitable phonon.

Finally, the probability that an electron transition takes place depends

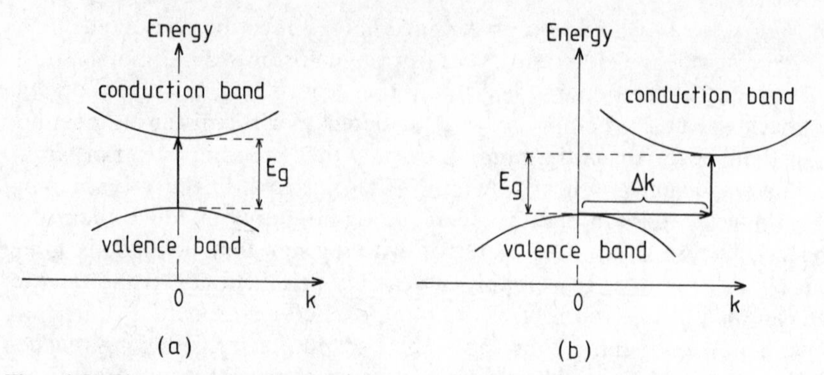

Fig. 4.5. The arrows indicate direct (a) and indirect, phonon-assisted (b) transitions of valence electrons into the conduction band after receiving energy larger than or equal to the band gap energy E_g from a collision with a photon. Δk is the momentum difference supplied by the vibrational wave (phonon).

on the form of the wave functions of the initial and final quantum states, i and f, respectively. In texts on quantum mechanics we would find an expression for transition probability per unit time in the form (the so-called Fermi Golden Rule)

$$w_{if} \sim \left| \int \psi_i^* \, ex \, \psi_f \, \mathrm{d}x \right|^2 \rho(\hbar\omega - E_{\mathrm{th}}), \tag{4.11}$$

where ρ is the density of final states the electron can go to, measured from the threshold value E_{th} of energy for the process (e.g. E_g); $\hbar\omega$ is the photon energy; and ex is the classical dipole representing the electromagnetic field of the light beam. Equation (4.11) describes a quantum-mechanical process analogous to that exhibited by an emitting Hertz dipole for which the rate of energy dissipation is proportional to $|ex|^2$. As usual, the quantum expression is obtained by taking an 'expectation value' of the classical term. If the transitions occur between discrete non-degenerate atomic levels, the density of states function ρ would be reduced to factor 1 when $\hbar\omega$ equals the difference between the relevant atomic levels and to zero when it does not. The Golden Rule provides a link between electronic structure of solids (i.e. the wave functions and energies) and the empirical world of facts, since experiments do not measure ψ or E_k but rather the transition probabilities and energies. Clearly, the content of eqn (4.11) is not trivial. A glance at the integral tells us that only for certain ψ_i, ψ_f is a transition possible. For example, if the wave functions of both the initial and final quantum states are even functions of x, then the integral vanishes and w_{if} is zero. Such general statements are called selection rules.

4.4. Impurity levels

How would the electronic structure of a crystal change if a small number of host atoms were replaced by foreign species? For example, let us replace $10^{15}\,\mathrm{cm}^{-3}$ silicon atoms in a silicon crystal by phosphorus. Since the average distance between phosphorus impurities is large, we can consider each phosphorus atom as an isolated impurity. Phosphorus is very similar to silicon but possesses five valence electrons, not four. If we assume that in the region around a P atom the valence band is not significantly changed, then we have four empty slots vacated by the removed silicon valence electrons. We can accommodate there four of the five phosphorus valence electrons. The next available level is at the bottom of the conduction band. If we place the remaining electron there, its kinetic energy will be $\hbar^2k^2/2m_e^*$. Except for a small region near the phosphorus nucleus, this electron will see the Coulomb potential $-(Z_P - Z_{Si})e/(4\pi\epsilon_0\epsilon r)$. This is because the nuclear charge at the phosphorus site (at $r = 0$) is shielded by the electrons in the atomic core and in the valence band, except for a charge $(Z_P - Z_{Si})e$, where Z_P, Z_{Si} are the atomic numbers of phosphorus and silicon atoms, respectively. Since $Z_P - Z_{Si} = 1$, the electron sees a potential that is the same as the potential for the hydrogen problem. The only difference in the present case is that the static dielectric constant $\epsilon(\mathrm{Si}) = 12$. Also, the electron effective mass is $m^* \approx 0.3m$ in this case. Therefore, the electron energy measured from the bottom of the conduction band continuum is that of a hydrogen atom, $(-13.6\,\mathrm{eV}/n^2)$, with ε and m^*, i.e.

$$\Delta E_n = E_{\text{imp}} - E_c = -13.6\frac{m^*}{m}\frac{1}{\epsilon^2}\frac{1}{n^2} \quad (\mathrm{eV}), \tag{4.12}$$

where n is the principal quantum number familiar from Bohr's model. Substituting for ϵ, m^* we obtain for the ground state the binding energy $\Delta E_1 \approx -28\,\mathrm{meV}$. The effective radius of the ground state wave function is $a^* = a_0\epsilon/m^* = 40a_0$, where $a_0 = 0.529 \times 10^{-10}\,\mathrm{m}$ is the radius of the lowest hydrogenic orbit. We can proceed to compute states lying closer to the conduction band edge, with $n = 2, 3, \ldots$.

It is clear that the above prescription may be used to predict new energy levels in the forbidden gap due to other impurities. We only have to know the host crystal parameters ϵ, m^* and the atomic number of the atoms that are being replaced. If the difference in Z is negative, we are short of valence electrons, so that by replacing silicon with, say, gallium, we are left with a hole at the top of the valence band that sees a Coulomb potential attractive to holes (repulsive to electrons). Consequently, we obtain new impurity levels above the valence band edge. The

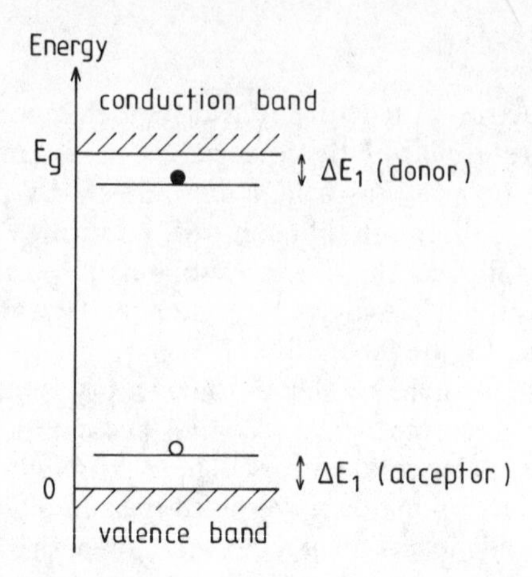

Fig. 4.6. The new energy levels introduced into the forbidden gap by donor and acceptor impurities such as phosphorus and gallium in silicon. ΔE_1 is the binding energy of the impurity measured from the corresponding bulk band edge and calculated from the hydrogenic formula in eqn (4.12).

phosphorus-like impurities are called *donors,* since they contribute 'extra' electrons. The gallium-like impurities are called *acceptors* (Fig. 4.6).

Impurities such as phosphorus or gallium in silicon are called *shallow* impurities because the energy levels introduced into the forbidden gap by them are such that $|\Delta E_1| \ll E_g$. Not all foreign elements can be described in this manner. When the foreign atom is very different from the host atom that is being exchanged, the details of the short-range part of the atomic potentials become important. Then the impurity level can lie anywhere in the forbidden gap and it is called a *deep* level. The wave functions associated with deep levels are well localized at the impurity site. This is in contrast with wave functions of shallow donors and acceptors, which have radii much larger than the nearest-neighbour distance between host atoms in the lattice. Deep levels are also introduced into the forbidden gap by foreign atoms sitting at interstitial positions in the lattice, or by the absence of an atom from a lattice site (interstitials and vacancies, resp.). Dislocations (missing half-planes of host atoms) may also give rise to such localized levels in the gap.

When the concentration of impurities become very high ($>10^{18}\,\mathrm{cm}^{-3}$), their wave functions begin to overlap significantly and electrons can hop from one impurity to another. Also, clusters of two or more coupled impurities are likely, so that new levels in the gap arise. The isolated impurity levels are then be broadened into bands in the same way as in periodic structures discussed in Chapters 2 and 3.

Finally, let us consider a special 'particle' called an *exciton* whose presence often provides a means of characterizing the optical properties of crystals. This particle is formed when an electron at the bottom of the conduction band of mass m_e^* and a hole of mass m_h^* meet and attract each other via the Coulomb force. Since the electron carries a negative charge, and the hole is positive, the interaction is formally similar to that of a donor electron and the extra proton in the impurity nucleus. Just as in the hydrogen atom problem, where the reduced mass μ for the proton–electron rotor is $\mu^{-1} = M^{-1}(\text{proton}) + m^{-1}(\text{electron}) \approx m^{-1}(M \gg m)$, we can define a reduced effective exciton mass m_{ex}^* from $(m_{\text{ex}}^*)^{-1} = (m_e^*)^{-1} + (m_h^*)^{-1}$. We end up with a hydrogenic Schrödinger equation whose solutions are

$$E_n^{\text{ex}} = -13.6 \frac{m_{\text{ex}}^*}{m} \frac{1}{\epsilon^2 n^2} (\text{eV}), \tag{14.12a}$$

and ground state wave function with an effective radius of $(\epsilon / m_{\text{ex}}^*)$ 0.529×10^{-10} m. The exciton has a short lifetime of about 10^{-9} s, after which the electron recombines with the hole and emits a photon of energy $E_g - |E_1^{\text{ex}}|$. Free excitons can travel in the crystal and be captured by impurities. This alters the exciton binding energy and its lifetime.

4.5. p–n Junctions

p–n Junctions are formed when a semiconductor crystal is doped with donors and acceptors as shown in Fig. 4.7. The donor electrons near the conduction band edge leave their sites and jump into the lower-lying hole (empty) acceptor levels on the other side of the junction. This redistribution of charge occurs only in the region of the junction and establishes an electric field that eventually prevents further charge transfer taking place. The region near the junction that is depleted of mobile carriers and contains only fixed chages due to positively charged donor atoms on the n-side and negatively charged acceptors on the p-side is called the depletion region. In order to characterize the junction, we must evaluate the magnitude of the potential difference created by the charge redistribution, the contact potential $e\phi$ of Fig. 4.7, and the width of the space-charge region l. We know from the preceding section that the concentration of electrons in the conduction band in an n-type material is

$$n_{\text{n}} = \text{const.} \exp[-(E(\text{n}) - E_F)/k_B T]. \tag{4.13}$$

Similarly, the concentration of free electrons in the conduction band of a p-type material is

$$n_{\text{p}} = \text{const.} \exp[-(E(\text{p}) - E_F)/k_B T], \tag{4.14}$$

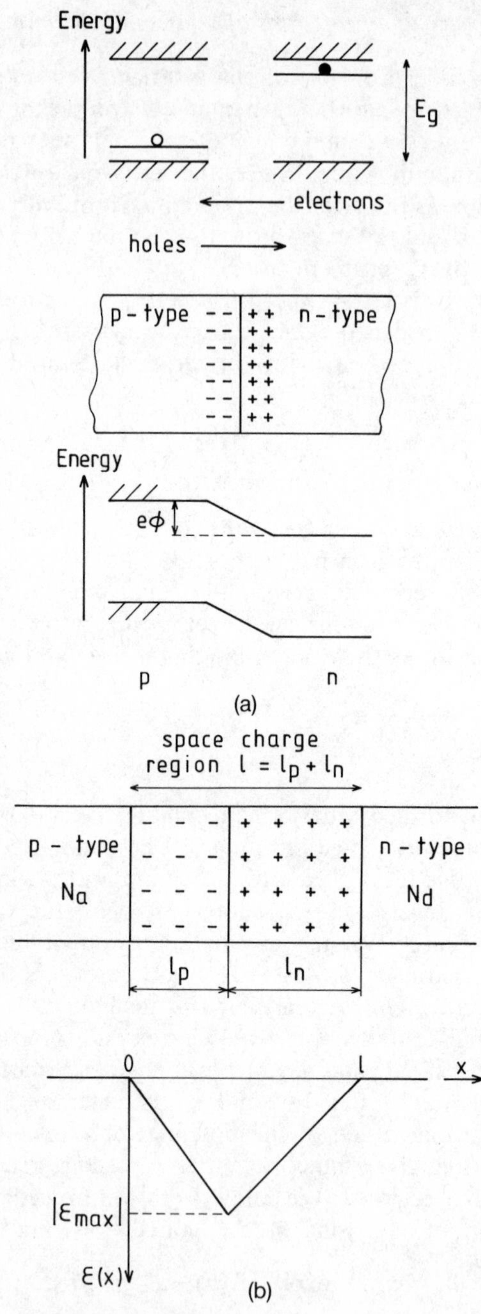

Fig. 4.7. (a) A semiconductor doped with donors (n-type) and acceptors (p-type) in the manner shown here is said to form a p–n junction; $e\phi$ is the contact potential characteristic of this junction. (b) The resulting profile of the electric field associated with $e\phi$ and reflecting the width of the depletion layer $l = l_n + l_p$.

where $E(\text{n})$, $E(\text{p})$ indicate the position of the conduction band edges in n- and p-type materials, respectively. The contact potential is

$$e\phi = E(\text{p}) - E(\text{n}), \tag{4.15}$$

and from eqns (4.13) and (4.14) we get the ratio n_n/n_p and ϕ:

$$\phi = \frac{k_B T}{e} \ln\left(\frac{n_\text{n}}{n_\text{p}}\right). \tag{4.16}$$

Note that E_F must be constant throughout both n- and p-type parts and, since it depends only on the total (average) electron density, it will reflect the concentration of donors and acceptors in the crystal. However, thanks to the law of mass action, we do not have to evaluate E_F explicitly.

From the law of mass action we know that $np = n_i^2$. This must be valid simultaneously in both n- and p-type parts, i.e.

$$n_\text{n} p_\text{n} = n_i^2, \qquad n_\text{p} p_\text{p} = n_i^2. \tag{4.17}$$

We need to know n_n/n_p in order to find ϕ. From eqn (4.17) we get

$$\frac{n_\text{n}}{n_\text{p}} = \frac{n_\text{n} p_\text{p}}{n_i^2} \tag{4.18}$$

At room temperature, $k_B T \sim 25$ meV, which is comparable to the depth (ΔE_1) of the donor and acceptor impurity levels eqn (4.12). Hence we can often simplify the calculation of n_n/n_p by making the assumption that all donors and all acceptors are ionized, i.e. $n_\text{n} \approx N_d$, $p_\text{p} \approx N_a$, where N_d, N_a are the donor and acceptor concentrations, respectively.

Let us now evaluate the width (l) of the depletion region in the approximation developed above (i.e. idealized depletion region with no free carriers, all impurities ionized). Consider the structure in Fig. 4.7. The charge neutrality condition that links all variables introduced in this figure is ($l = l_\text{p} + l_\text{n}$)

$$N_a l_\text{p} = N_d l_\text{n}. \tag{4.19}$$

Let us apply Gauss' law to the p-side of the junction:

$$\frac{\text{d}\mathscr{E}}{\text{d}x} = -\frac{eN_a}{\epsilon_0 \epsilon}, \tag{4.20}$$

where $\mathscr{E}$ is the electric field. Integrating, we get

$$\mathscr{E} = -\frac{eN_a}{\epsilon_0 \epsilon} x + B. \tag{4.21}$$

Since $\mathscr{E} = 0$ at $x = 0$, the constant $B = 0$, and the maximum value of the field is

$$|\mathscr{E}_{\max}| = \frac{eN_a l_p}{\epsilon_0 \epsilon}. \tag{4.22}$$

Using the same procedure for the other side of the junction, we obtain the condition

$$|\mathscr{E}_{\max}| = \frac{eN_d l_n}{\epsilon_0 \epsilon}. \tag{4.23}$$

The potential difference ϕ is

$$\phi = \int_0^{l_n + l_p} \mathscr{E}(x)\, dx = \frac{e}{2\epsilon_0 \epsilon}(N_a l_p^2 + N_d l_n^2) = \frac{l}{2}|\mathscr{E}_{\max}|. \tag{4.24}$$

We can now recall eqn (4.16) with eqn (4.18), to write

$$\phi = \frac{k_B T}{e} \ln\left(\frac{N_a N_d}{n_i^2}\right). \tag{4.25}$$

Using eqns (4.19), (4.24) and (4.25), we find $l(=l_n + l_p)$ from

$$\begin{aligned} l_n &= \left(\phi \frac{2\epsilon_0 \epsilon}{eN_d} \frac{N_a}{N_a + N_d}\right)^{1/2}, \\ l_p &= \left(\phi \frac{2\epsilon_0 \epsilon}{eN_a} \frac{N_d}{N_a + N_d}\right)^{1/2}. \end{aligned} \tag{4.26}$$

We can use l to determine the junction capacitance $C = A\epsilon_0 \epsilon / l$, where A is the junction area.

4.6. Diffusion. Einstein's relation

The motion of carriers (electrons and holes) by diffusion occurs whenever the carrier distribution is non-uniform. In this process carriers are moving to a more uniform distribution. We can imagine that each carrier in the non-uniform ensemble moves randomly in all directions, so that the number of carriers crossing from high-density regions into low-density ones is larger than the number of those going the opposite way. Clearly, this process depends only on the concentration gradient and does not involve the Coulomb (electrical) forces. The diffusion flux of carriers in the x direction is proportional to dn/dx, where n is the carrier concentration. The diffusion electric current density is therefore

$$j_{\text{diff}} = -qD \frac{dn}{dx}, \tag{4.27}$$

where D is the diffusion constant and q is the charge of the diffusing carrier (electrons, holes). The concentration n will change with time as carriers diffuse from regions of large n. This change can be expressed as a continuity equation:

$$\frac{\partial n}{\partial t} = -\text{grad}(j_{\text{diff}}/q), \tag{4.28}$$

If an electric field $\mathscr{E}$ is applied to the sample, we must include its contribution to the current density j, e.g. for electrons ($j = j_d + j_{\text{diff}}$)

$$j = e\mu n\mathscr{E} + eD\frac{\mathrm{d}n}{\mathrm{d}x}, \tag{4.29}$$

where μ is the electron mobility (μ is defined in Section 4.7). The equilibrium electron density at T is (Boltzmann's law) $n = \text{const.}\ \exp(-E/k_BT)$, where $E = -eV(x)$ is the excitation energy, which reflects the total charge distribution in the crystal (potential $V(x)$) and is therefore a function of x. At equilibrium, $j = 0$ so that

$$e\mu n\mathscr{E} + eD\frac{\mathrm{d}n}{\mathrm{d}x} = 0. \tag{4.30}$$

Substituting for n and differentiating gives us

$$\mu\mathscr{E} + D\frac{e}{k_BT}\frac{\mathrm{d}V}{\mathrm{d}x} = 0. \tag{4.31}$$

The only way to satisfy this equation is to equate $\mathscr{E}$ with $-\mathrm{d}V/\mathrm{d}x$ and set

$$D = \frac{k_BT}{e}\mu. \tag{4.32}$$

This is the so-called Einstein relation, which links mobility μ to diffusion processes.

4.7. Electron motion in an electric field. Relaxation time. Recombination

The electron current per area is the current density $j = -nev$, where n is the electron density, $-e$ is the electron charge, and v is velocity. In the absence of an external electric field, the average velocity is zero (unless the electron distribution in the crystal is non-uniform; see Section 4.6). In an external electric field $\mathscr{E}$, electrons acquire in time t a velocity $-e\mathscr{E}t/m^*$. If their initial velocity (at $t = 0$) is v_0, then

$$v = v_0 - \frac{e\mathscr{E}t}{m^*}, \tag{4.33}$$

where m^* is the effective mass of the corresponding band extremum. An analogous expression exists for a hole current. The force $-e\mathscr{E} = m^* \, dv/dt$ accelerates the electron until it collides with an impurity or a lattice wave (phonon) and loses its energy (relaxes from v to v_0). This happens on average once in time τ; τ is called the relaxation time. The equilibrium (drift) velocity acquired in the electric field is therefore

$$v_d = -\frac{e\mathscr{E}\tau}{m^*}, \tag{4.34}$$

and the steady state current density is

$$j_d = -nev_d. \tag{4.35}$$

The drift velocity per electric field is the mobility μ:

$$\mu = \left|\frac{v_\mathrm{d}}{\mathscr{E}}\right| = \frac{e\tau}{m^*}. \tag{4.36}$$

We can express τ in terms of observable quantities such as resistivity β given by $j_d = \mathscr{E}/\beta$. From eqns (4.34)–(4.36) we obtain

$$\tau = \frac{m^*}{\beta n e^2}. \tag{4.37}$$

During time τ the electron overcomes a distance $l = v\tau$; l is the *mean free path.* It is the average distance between subsequent collisions. Examples of electron and hole mobilities in semiconductors and the relevant band structure parameters are given in Table 4.1.

The above relations were derived for free-electron gas and adapted for application in semiconductors by approximating the curvature of the conduction or valence band in terms of the effective mass m^*, i.e. $E_k = \hbar^2k^2/(2m^*)$. When these transport equations are applied to simple metals, we obtain ($m \approx m^*$)

$$v_0 = \left.\frac{\mathrm{d}\omega}{\mathrm{d}k}\right|_{k=k_F} = \left.\frac{1}{\hbar}\frac{\mathrm{d}E_k}{\mathrm{d}k}\right|_{k=k_F} = \frac{\hbar}{m}k_F. \tag{4.38}$$

Table 4.1. Summary of band structures parameters for simple room-temperature junctions, impurity and transport calculations

	E_g (eV)	m_e^*/m	m_h^*/m	μ_e (cm^2 V^{-1} s^{-1})	μ_h(cm^2 V^{-1} s^{-1})	NEV[a]	ϵ[b]	n_i (cm^{-3})
Si	1.12	0.33	0.55	1500	480	6	12	1.5×10^{10}
Ge	0.66	0.22	0.29	3900	1900	4	16	2.4×10^{13}
GaAs	1.42	0.067	0.62	8500	400	1	13.18	1.8×10^6

[a]NEV is the number of equivalent conduction edge valleys (see Chapter 5).
[b]ϵ is the static dielectric constant: ϵ (AlAs) = 10.06 and $\epsilon(SiO_2) = 3.9$.

The typical value of resistivity β is 10^{-7} (ohm m) for $\tau \approx 10^{-14}$ s. For weak and moderate fields $v_d \ll v_0 \approx 10^6\,\mathrm{m\,s^{-1}}$; k_F is the Fermi wave vector ($E_F = \hbar^2 k_F^2/2m$). In semiconductors, we obtain τ of the same order of magnitude from eqn (4.37). Although resistivity is larger, the number of electrons available (n) is much smaller. However, the electron group velocity at the band edge is zero, since the electron wave function there is a standing wave; v_0 is therefore just the thermal velocity. In classical thermodynamics, we find that at temperature T a free particle has energy $\frac{1}{2}k_B T$ per degree of freedom. Hence, we obtain

$$v_{\text{thermal}} = \sqrt{\frac{3k_B T}{m^*}} \quad (\approx 10^4\,\mathrm{m\,s^{-1}} \text{ at } 300\,\mathrm{K}). \tag{4.39}$$

We have argued that the relaxation time measures the frequency of collisions the electron suffers along its path. We can use the experimental values of τ to compute $l = v\tau$ and find l of order 100 Å. This is consistent with the view expressed earlier that an electron driven by the electric field does not lose energy by colliding with individual atoms in the lattice (which are separated by a few angstroms). The collisions are more likely to occur at larger wavelengths that characterize the vibrational waves (phonons). At higher temperatures, there are more vigorous vibrations, so that τ decreases and so does conductivity. It is assumed that the time it takes to make a collision is short compared to the time (τ) between collisions. If it were not, our model and transport equations would collapse. We also tacitly assume that the cooling of the accelerated electrons is fast enough that there are empty states in the conduction band at higher energies (above the semiconductor band edge) for the electrons to go to. Both these assumptions become suspect when the electric field is very large. Eventually, the drift velocity reaches the maximum thermal velocity, so that the electrons can no longer follow any increase in the field. We say that they reach the *saturation velocity*.

At low temperatures, the lattice vibrations are suppressed and the electron energy is carried away mainly by collisions with impurities. Thus, the low-temperature mobility depends critically on the impurity concentration along the electron path.

In our account of the collision processes we have assumed that electrons (or holes) instantaneously lose the energy and momentum gained in the accelerating field, so that they all return into the conduction currents. However, in some cases the carriers are taken away from the conduction process. One such mechanisms was encountered at the beginning of Section 4.3, where we considered an electron in the conduction band making a jump into an empty (hole) state in the valence band and disposing of the excess energy E_g by emitting a photon of frequency ω, where $\hbar\omega = E_g$. Such a radiative recombination event may occur between an electron in an impurity level and a hole travelling along (a hole capture), or in the anihilation of an exciton. In each case the energy of the emitted light is such that the energy of the total system is conserved. In some cases the radiative transition may be phonon-assisted, as in the case of indirect band-to-band transitions considered in Section 4.3. When all energy is dissipated by generating lattice waves (phonons), we say that the recombination or capture is non-radiative. For example, an electron in the conduction band may be captured by an ionized donor, e.g., a donor whose impurity levels are empty and available for an incoming electron to settle down in one of them. The approaching electron can lose its energy in steps. We know that the typical phonon energy is in the range of 10–40 meV. This is the order of separation between the hydrogenic impurity levels lying just below the conduction band edge. Thus, our electron can emit one phonon to climb down from the conduction band into one of the higher hydrogenic levels. Then it emits another phonon to descend closer to the ground state, and so on, until finally the lowest level is reached. The probability that a particular recombination (capture) event occurs determines the time an electron can stay in a given state in the conduction band. It also determines, together with the relaxation time τ for collisions, the time the electron is available to carry current and the speed with which the current travels.

Problems

4.1. Use the room temperature data given in Chapter 4 to calculate the ratio of the intrinsic carrier density in germanium and silicon.

4.2. Consider a silicon crystal at room temperature T containing 10^{16} donors per cm^3 with binding energy $\Delta E_1 < k_B T$. Estimate the Fermi energy of this system.

4.3. Consider a vibrating one-dimensional lattice of atoms A and B shown in Fig. 4.4a. Use Hooke's law (displacement is proportional to applied force, with proportionality constant α) and carry out a classical equation of motion calculation to generate the normal mode dispersion curves that are shown in Fig. 4.4b.

4.4. Show that the absorption spectrum of a nearly-free-electron insulator is characterized by strong peaks at photon energies corresponding to direct (vertical) transitions across the gap at the Brillouin zone boundaries.

4.5. Estimate the depth of the ground state of Zn^- and Se^+ impurities in silicon and germanium crystals.

4.6. Calculate the contact potential of a p–n junction in Si and GaAs at room temperature. The concentration of donors and acceptors is $N_d = N_a = 10^{16}\ cm^{-3}$.

4.7. Calculate the width of the depletion layer of an abrupt p–n junction in silicon at room temperature. The n-side is doped with 5×10^{14} cm^{-3} donors and the p-side with 5×10^{19} cm^{-3} acceptors.

4.8. A p-type sample of width 1.5 μm containing 10^{15} cm^{-3} acceptors is injected with electrons so that a uniform gradient of electron concentration is formed from zero to $N_e = 10^{14}$ cm^{-3}. Determine what electric field must be applied to generate a drift electron current that is exactly equal to the diffusion current density ($T = 300$ K).

4.9. Compare electron conductivity in a simple metal and in a semiconductor at room temperature. Assume that the relaxation times and m^* are the same in both materials.

4.10. A sample of silicon crystal is illuminated with light of wavelength λ at $T = 300$ K. What is the threshold value of λ in angstroms at which electron–hole pairs are created? If the carrier lifetime in this sample is 1 μs, how long will it take for the electron pair concentration to be reduced by 90%.

5

Electronic structure of technologically important semiconductors

In Chapters 2 and 3, we developed a simple one-dimensional picture of the electronic structure of an insulator. We found that it is profitable to think of the interaction between electrons and the lattice of positive ions in terms of a classical diffraction process. The key features in the electronic structure are then naturally depicted in a dispersion diagram of the electron energy versus wave vector **k**. They occur at special values of **k** that we called the Brillouin zone boundary, which can be obtained directly from the Bragg condition. The electronic structure of a one-dimensional insulator consists of bands of allowed energies separated by forbidden gaps. The most important part of this band structure lies at the fundamental gap, i.e. in the region of energies dividing the uppermost occupied states at the top of the valence band from the lowest empty states at the bottom of the conduction band. In the spirit of the nearly-free-electron model, we expect the free-electron Fermi energy to lie in the middle of this fundamental gap. Of course, real solids are three-dimensional objects and although our one-dimensional theory is quite adequate for providing us with the necessary concepts, it cannot furnish any details of the electronic structure derived from the crystal symmetry. We shall see later than such 'details' often represent an indispensable factor in our appreciation of useful physical properties of semiconductor materials.

The simplest way to extend our theory to three dimensions is to assume that the electronic structure is the same no matter which direction in the wave vector space we choose. The band structure diagram is then formally the same as that obtained for our one-dimensional solid, with the fundamental gap being constant everywhere over the surface of the Fermi sphere of radius k_F (Fig. 5.1). Most technologically important semiconductors are either diamond-like or zinc-blende structures, shown in Fig. 5.2. It is clear from a glance at this figure that the crystal structure is quite complicated and that the electronic structure must be expected to vary depending on the direction along which we choose to

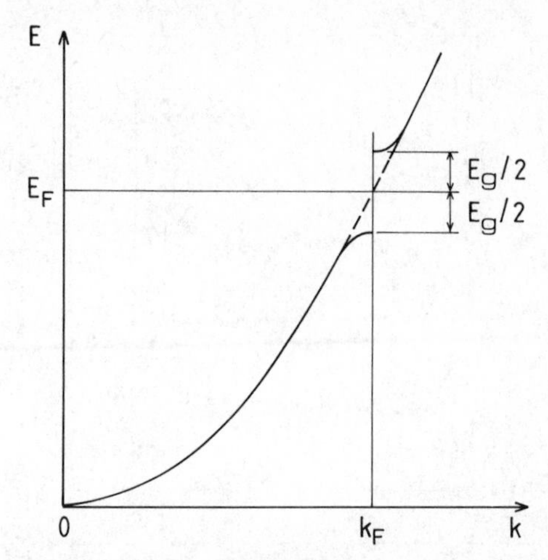

Fig. 5.1. The electronic structure of a model (isotropic) semiconductor. E_F and k_F are the free-electron Fermi energy and wave vector, respectively. E_g is the forbidden gap separating the occupied and empty states.

plot energy as a function of **k**. In order to account for the symmetry properties of the crystal, we must extend the procedure we adopted in Chapter 2 in the calculation of the electronic structure of a one-dimensional model solid. However, such calculations lie outside the scope of this course. It is only the results that interest us here.

The electronic band structures of some important semiconductors are shown in Fig. 5.3. It is essential that we know how to "read" such

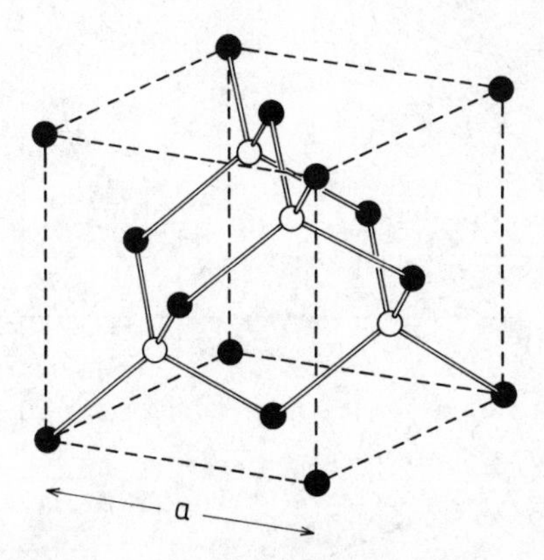

Fig. 5.2. The conventional unit cell showing a cube of length a (the lattice constant) and the arrangement of atoms corresponding to a compound semiconductor such as GaAs (for which the open circles would represent gallium atoms and the solid ones arsenic atoms).

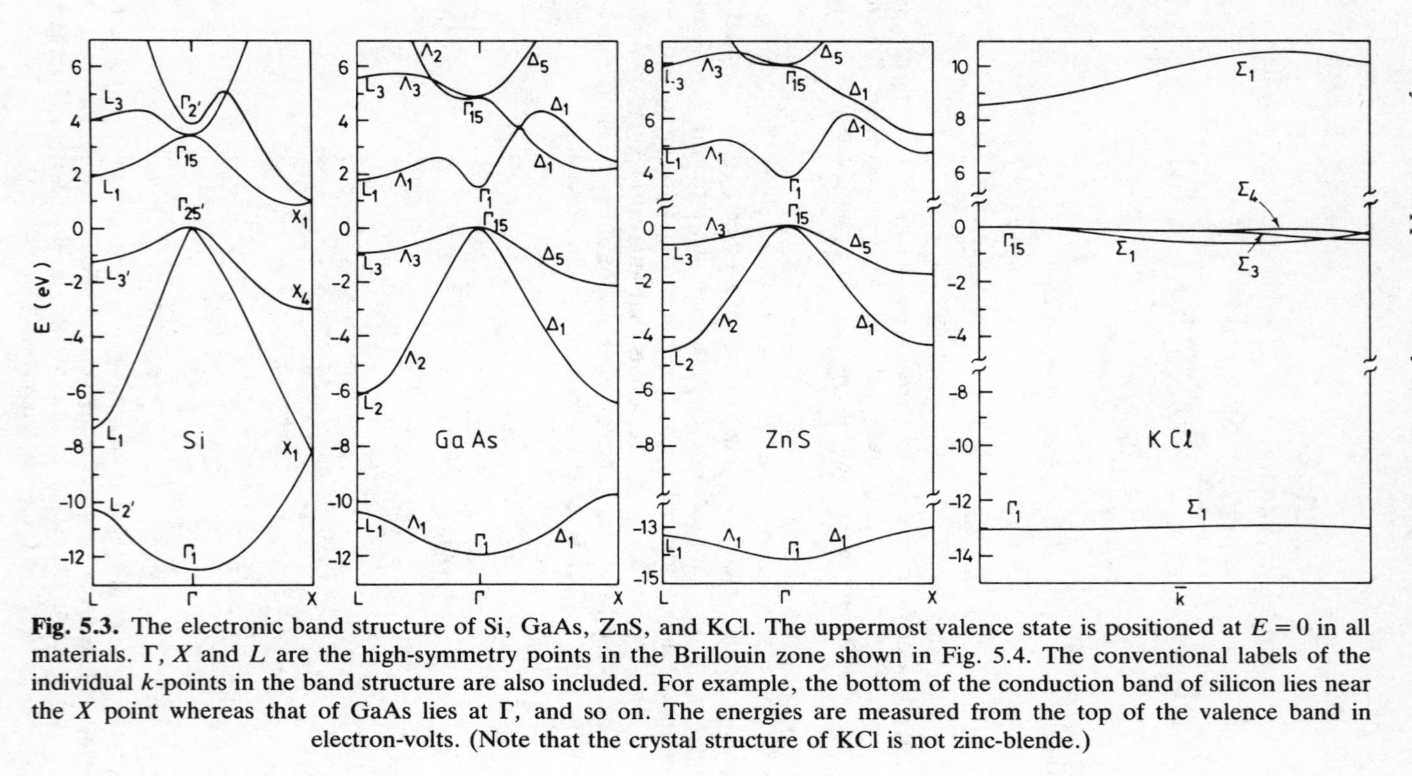

Fig. 5.3. The electronic band structure of Si, GaAs, ZnS, and KCl. The uppermost valence state is positioned at $E = 0$ in all materials. Γ, X and L are the high-symmetry points in the Brillouin zone shown in Fig. 5.4. The conventional labels of the individual k-points in the band structure are also included. For example, the bottom of the conduction band of silicon lies near the X point whereas that of GaAs lies at Γ, and so on. The energies are measured from the top of the valence band in electron-volts. (Note that the crystal structure of KCl is not zinc-blende.)

diagrams. First of all, it is useful to know that we normally view the crystal structure of Fig. 5.2 as a face-centred cubic lattice, with each lattice point (see Chapter 1, Fig. 1.4) representing two atoms. In the case of elemental semiconductors (from the fourth column of the Periodic Table), these two atoms are the same species (e.g. Si). In the case of compound semiconductors such as GaAs (a so-called III–V compound material because its smallest or primitive unit cell used as a building block here consists of one atom from the third column (Ga) and one from the fifth (As)) or ZnSe (a II–VI compound semiconductor), each cubic lattice points represents a pair of Ga–As and Zn–Se, respectively. The lattice constant a normally quoted is that of the conventional unit cell, i.e. the length of the cube shown in Fig. 5.2. Thus the nearest-neighbour distance is $a\sqrt{3}/4$ and there are eight atoms (32 electrons) per volume a^3. The three-dimensional Brillouin zone that corresponds to this crystal structure is shown in Fig. 5.4. The details of its shape are not important here. We just want to be familiar with the language used to refer to the properties of band structures. For instance, it is customary to associate the most important (high-symmetry) points with capital letters Γ, X, L etc. These points lie at the Brillouin zone centre and at the boundaries, along the high-symmetry lines that are normally referred to in terms of the Miller indices. For example, the Γ–X direction is the main cubic axis $\langle 001 \rangle$—if X is the point in **k**-space whose x and y components are zero.

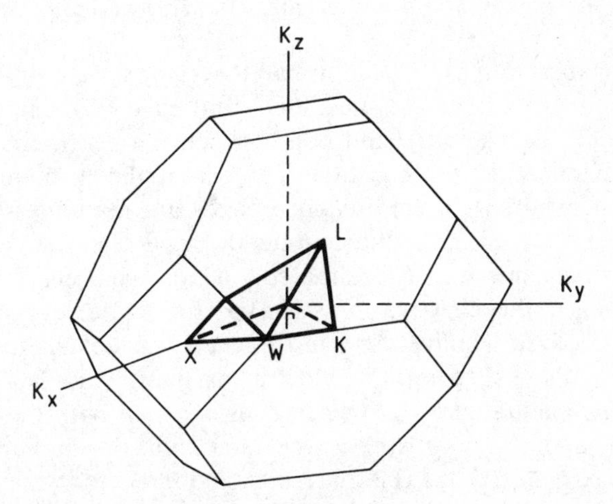

Fig. 5.4. The first (the so-called reduced) Brillouin zone of the face-centred cubic lattice. The points of high symmetry such as Γ, X, and L are indicated. The heavy line marks the so called irreducible segment, i.e., the smallest section of he zone that can be used to generate the Brillouin zone by applying to it the symmetry operations under which the face-centered cubic lattice is invariant.

These are six such directions, but since they are equivalent we only show a plot of the electronic structure along one of them. The $\Gamma - L$ direction lies along the line connecting the nearest neighbours in the lattice, i.e. along the $\langle 111 \rangle$ direction. Since we associate each lattice point of the face-centred cubic lattice with two atoms, we have eight electrons in the primitive unit cell and we therefore need four bands to accommodate all valence electrons (one band can take two electrons per atom). Thus, the fundamental forbidden gap separating occupied and empty states always lies between the fourth and the fifth bands. We can now inspect the diagrams shown in Fig. 5.3. The uppermost occupied state lies at the centre of the Brillouin zone (at the Γ point). The energy scale is chosen so that the top of the valence band lies at zero energy. (Notice that there are only three bands shown below $E = 0$, not four. In fact the top curve represents two bands whose energies happen to be the same along the high-symmetry direction shown in Fig. 5.3.) We can see that there are two curves converging to the top end of the valence band at Γ. In particular, the upper curve is rather flat. Since a flat band means that the corresponding effective mass (which is inversely proportional to the second derivative of E with respect to $\mathbf{k}$) is large, it is referred to as the heavy mass or heavy hole band. The other band, whose curvature is larger, and which moves rapidly down in energy as we go away from the Γ point, is called the light mass or light hole band. The light hole effective mass is typically about one-tenth of the free-electron mass. Examples of the value of the heavy hole effective mass in semiconductors are given in Table 5.1.

Let us now turn to the conduction band. Again, we are only interested in the description of the states near the band edge. We can see, first of all, that unlike the valence band edge (which always occurs at Γ), the conduction band edge in some materials lies at the X point. It is also apparent that, whereas the top valence bands are featureless monotonic functions of the wave vector, the conduction band is warped and exhibits distinct secondary minima. At Γ, the conduction band curvature is large and consequently the effective mass is expected to be very small. Some typical values of the conduction band effective masses are presented in Table 5.1. We also list there the dielectric constant ϵ, the lattice constant a and the fundamental gap E_g^f. The crystals in which both the conduction and valence band extrema occur at the same point in the Brillouin zone (e.g., at Γ) are called *direct gap* materials, and those for which this is not so are called *indirect gap* materials. For instance, silicon and GaP are indirect gap materials. GaAs, InP, ZnSe, ZnS, CdTe (and others) are direct gap materials. The energy separation between the principal and secondary minima in the conduction band is also an important band structure parameter. We can see that the separation between the Γ and

Table 5.1.

	a (Å)[1]	ϵ^{o2}	E_g (eV)[3]	E_g^f (eV)[3]	m_h^{*4}	m_e^{*4}
Si	5.43	12	5.00	†1.13	0.55	0.33
Ge	5.65	16	4.035	†0.76	0.29	0.22
α-Sn	6.47	24	2.66	0.0	0.5	0.02
AlAs	5.66	8.16	5.82	†2.36	0.76	0.26
GaAs	5.65	10.9	4.97	1.52	0.62	0.067
GaSb	6.10	14.4	3.80	0.81	0.49	0.045
GaP	5.44	9.1	5.81	†2.38	0.60	0.17
InP	5.86	9.6	5.04	1.37	0.85	0.077
InAs	6.05	12.3	4.20	0.42	0.60	0.024
InSb	6.47	15.7	3.33	0.24	0.47	0.0137
ZnSe	5.65	5.9	7.06	2.82	1.44	0.17
ZnTe	6.10	7.3	5.55	2.39	1.27	0.16
CdTe	6.48	7.2	5.11	1.60	1.38	0.096
HgTe	6.48	9.3	4.42	−0.3	0.3	0.026

[1] a is the lattice constant in angströms.
[2] The optical dielectric constant ϵ^o, not to be confused with the static dielectric constant in Table 4.1.
[3] The average direct gap E_g obtained from eqn (5.2), and the fundamental gap E_g^f ($T \simeq 4$ K).,† indicates that the fundamental gap is indirect.
[4] m_h^* and m_e^* are the uppermost (heavy hole) valence band and the lowest conduction band effective masses in units of free electron mass, respectively. The band structure at the X and L minima is, in fact, strongly anisotropic. The value of the effective mass at these points given in this table for the indirect gap material is an 'average' value that would be obtained, for example, in a measurement of mobility.

X, and between the Γ and L minima varies from a few millielectronvolts (meV) in, say, germanium to about 1 eV in some II–VI compounds. The significance of this parameter becomes apparent if we consider electron transport. For example, an electron at the bottom of the conduction band of GaAs lies in the Γ valley. In a transport process, it is accelerated by an external electric field and gains some kinetic energy that lifts it temporarily above the band edge. The probability that the electron escapes from the Γ valley into the secondary X or L valleys decreases exponentially with increasing energy separation between the two states (i.e. the Γ and X minima) in question. The magnitude of the external field that can be usefully employed must be limited by the magnitude of the separation between the principal and secondary valleys. The positions of the secondary minima for several III–V semiconductors are shown in

Table 5.2. Forbidden gaps (in eV): $\Gamma-\Gamma$, $\Gamma-X$ and $\Gamma-L$ refer to the energy differences between the Γ, X and L conduction band minima and the top of the valence band at Γ, respectively ($T\simeq 4$ K)

	$\Gamma-\Gamma$	$\Gamma-X$	$\Gamma-L$
GaP	2.77	2.38	2.5
GaAs	1.52	1.98	1.81
InP	1.37	2.1	2.0
InAs	0.42	2.1	1.77
AlAs	3.05	2.36	2.9

Table 5.2. Note also that the band curvature and consequently the effective mass at the secondary valleys at the X and L points are quite different. For instance, in GaAs the effective mass at X is about five times larger than that at the Γ point (where $m^* = 0.067m$). Thus, an electron excited from Γ to X gains energy and at the same time its velocity is reduced, a situation that is equivalent to having a negative resistance circuit. The effective mass characteristic of the valley in which the mobile electron resides is one of the most important transport parameters. We shall invoke the many-valley character of the conduction band on a number of occasions in the chapters that follow.

The magnitude of the fundamental gap, the effective masses and the position of the principal and secondary minima are often referred to as optical and transport band structure parameters, because they play a key role in determining the observable properties of semiconductor materials. We shall see that these band structure parameters are also indispensable in the effort to engineer new artificial materials with tunable properties that can be exploited in device applications.

The most important optical parameter of a semiconductor material is the magnitude of the fundamental gap E_g^f. Photons whose energy is less than E_g^f cannot be absorbed. If we desire an efficient emitter or absorber of electromagnetic radiation at a particular wavelength, we must find in nature a material of suitable band gap. Since the momentum of the absorbed or emitted photon is normally negligibly small, momentum conservation requires that the optical transition be 'vertical', i.e. that the electron jump excited by the electromagnetic radiation occurs only between states associated with the same value of the wave vector in the reduced (first) Brillouin zone. Hence, indirect gap materials are not efficient light absorbers near the band gap energy because the valence and conduction band states at the edge have different wave vectors. We say that such transitions are forbidden because they do not satisfy the momentum conservation rule mentioned above. This further restricts the number of materials suitable for applications in optoelectronics.

The first attempts to 'tune' the optical properties of semiconductors led to the development of semiconductor alloys. For example, the band gap of GaAs is too small for light emission in the visible range. The band gap of GaP, on the other hand, is in the green portion of the visible band, but because GaP has an indirect gap this material is not an efficient emitter without the help of suitable dopants (nitrogen). An alloy of GaP and GaAs of well-chosen composition can ensure that the most important characteristics of GaAs are retained (i.e. the direct gap) while the magnitude of the gap is suitably increased. The simplest way of predicting what the electronic structure of the alloy might be is to assume that the alloy can be represented by a crystal with the relevant concentrations of arsenic and phosphorus. The band structure, the lattice constant, and other materials parameters are then obtained by linear interpolation between the values of the corresponding parameters for GaP and GaAs. For example, the lattice constant of an alloy with, say, 30% P ($x = 0.3$) can be calculated from the lattice constant of GaP and GaAs as follows:

$$a(\mathrm{GaP}_x\mathrm{As}_{1-x}) = a(\mathrm{GaP}) + [a(\mathrm{GaAs}) - a(\mathrm{GaP})](1-x). \qquad (5.1)$$

The assumption that the alloy is a perfect regular ordered crystal like GaAs or GaP means that we expect the distribution of phosphorus and arsenic to be strictly random, so that we can define an average crystal potential and band structure. Experimental results show that this approximation is quite good. In Figure 5.5, we can see the near-linear variation of the key band gaps (energy separations between the conduction band minima at Γ, X and L and the top of the valence band at Γ) as a function of the aluminium concentration x in $\mathrm{Ga}_{1-x}\mathrm{Al}_x\mathrm{As}$ obtained in an experimental study.

In an effort to optimize the applicability of semiconductor materials, ternary (e.g. $\mathrm{Ga}_{1-x}\mathrm{Al}_x\mathrm{As}$) as well as quaternary alloys (e.g. $\mathrm{Ga}_{1-x}\mathrm{In}_x\mathrm{As}_{1-y}\mathrm{P}_y$) have been developed. There are two variable parameters in the quaternaries and that permits a choice of band gap as well as the lattice constant. The latter is also an important facility because of the need to choose a suitable lattice-matched substrate material on which a high-quality alloy must by grown.

Let us now return to the collection of band diagrams in Fig. 5.3 with a view to recovering some of the basic trends we managed to predict from our studies of a one-dimensional model insulator in Chapters 2 and 3. First, recall the distinction we made between covalent materials such as

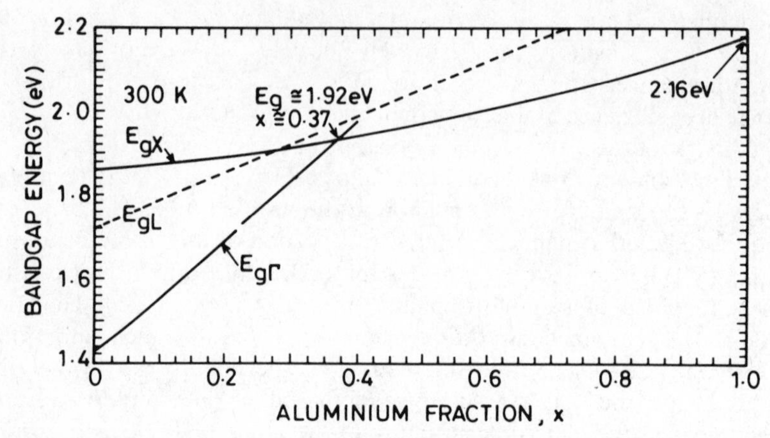

Fig. 5.5. Experimental data taken at room temperature and concerning the variation of the gaps of $Ga_{1-x}Al_xAs$ at the Γ, *X*, and *L* points, as a function of the aluminium concentration.

silicon and the ionic ones such as NaCl. We argued that in silicon the atoms are bonded together because of electron-sharing that leads to charge being transferred from the atoms to the space between them along the ⟨111⟩ directions. In NaCl, the outer electron is completely transferred from Na to Cl. Hence, in the former case we expect the wave functions to be plane wave-like, whereas in the latter they should resemble the wave functions of free atoms. We demonstrated that this difference in the physics of the formation of bonds is translated into clear band structure features; broad nearly-free-electron bands and small gaps in the covalent case, and narrow bands and large gaps in the ionic case. The band structures of the materials shown in Fig. 5.3 have been lined up so that the top of the valence band of each material lies at the same energy, to make the comparison easier. We can see that the trend predicted in the simple model is well accounted for.

We can also see that with the exception of the region near Γ, the direct (vertical) band gap in a given material is nearly constant. In fact, the deviation from the constant gap that is found in the region near the Γ point is not very significant when we realize that this region accounts for a small fraction of the volume of the Brillouin zone. This is well in keeping with the nearly-free-electron model of the forbidden gap that we proposed at the beginning of this chapter to describe the simplest three-dimensional isotropic insulator (Fig. 5.1). We can exploit this remarkable feature to establish a simple relation between the band structure (band gap) and the dielectric constant.

In general, the dielectric constant of a system such as a free atom, a molecule, or an insulator measures the polarizability of this system in an external optical field of a given frequency. We shall show in Chapter 9 (eqn (9.24)) that the optical dielectric constant can be expressed as

$$\epsilon^o = 1 + (\hbar\omega_p/\delta E)^2 \tag{5.2}$$

where $\omega_p = (Ne^2/\epsilon_0 m)^{1/2}$, and $\delta E = \hbar\omega_0$, where ω_0 is the characteristic frequency of the system. If, for example, we choose to study a classical oscillator, $\omega_0 = (\beta/m)^{1/2}$, where β is the force constant and m is the mass. In quantum mechanics, the ability of a system to exhibit polarization depends on the separation δE between the ground state and the first empty state to which the particle in question can be excited in response to the external field. Hence, in the case of our insulator in Fig. 5.1, we must replace δE by E_g. Since the empirical value of ϵ^o is known for most solids, and the plasma frequency ω_p depends only on electron density N, we can predict the magnitude of the gap shown in Fig. 5.1 from eqn (5.2). The result is given in Table 5.1. Although eqn (5.2) is a crude approximation, a glance at Fig. 5.3 suggests that the magnitude of the gap is well represented. For instance, in silicon the observed value of the band gap at X is 4.2 eV. We must remember that the gap obtained from ϵ^o in eqn (5.2) is really the gap of an idealized isotropic semiconductor shown in Fig. 5.1 and should be compared with the average over all the Brillouin zone of the direct gaps obtained in a full calculation. However, the conduction and valence bands shown in Fig. 5.3 are quite flat and the gap is constant over a large region around the X point, so that the band structure properties at X will dominate the value of ϵ^o obtained from eqn (5.2).

Summary

A realistic appreciation of the band structure of semiconductors can only be obtained from large calculations that lie outside the scope of this course. However, the result of such calculations can readily be understood in terms of the concepts established in simple one-dimensional models in preceding chapters. The most important features of the band structure are the positions of the valence and conduction band edges, and the structure of the conduction band, which exhibits several minima with distinctly different effective masses and energy separations. The uppermost occupied states at the top of the valence band always lie at the centre of the Brillouin zone (at the Γ point). There are heavy and light hole mass bands. Their energies are the same at Γ, but the two bands are very different away from Γ. The electronic structure parameters of semiconductor alloys can be obtained from those of the elemental and compound semiconductors by linear interpolation. These band structure parameters determine the optical and transport properties of semiconductors. A summary of the relevant parameters for some key materials is given in Tables 5.1 and 5.2.

Problems

5.1. Examine the position of atoms in the unit cube of Fig. 5.2.
(a) Use Table 5.1 to calculate the next-nearest distance between atoms in terms of the lattice constant *a* and compare it with the distance between nearest neighbours.
(b) Compare separations between atoms in materials consisting of atoms from the same row of the Periodic Table (e.g. Si, GaP).

5.2. Calculate the free-electron Fermi energy for Ge, GaAs, and ZnSe and compare the ratio of the forbidden gap and the Fermi energy for these materials.

5.3. Draw a diagram showing the positions of Si and GaAs atoms in (001), (111) and (110) planes. Assume that these planes represent a surface created by cutting the crystal into two halves and find how many equivalent surface planes there are in each case.

5.4. Use Table 5.1 to relate the dielectric constant and the observed and calculated forbidden gaps in materials consisting of atoms from the same row of the Periodic Table. List these materials in the order of increasing ionicity. Explain the result in simple qualitative terms.

5.5. The density of electron states was defined as the number of states in a unit energy interval. Inspect visually the band structures in Fig. 5.3 to identify the transition energies (gaps) at which a large density of states should occur. How does this affect optical properties of these materials?

5.6. Find a compound semiconductor or a semiconductor alloy such that the material is likely to be an efficient absorber of visible light in the red region of the spectrum.

5.7. Find the concentration of phosphorus in GaP_xAs_{1-x} such that the lowest states of the X and Γ valleys occur roughly at the same energy. What is the lattice constant of this alloy?

5.8. Use the results presented in Chapter 4 and in Table 5.1 to assess the depth of the impurity levels introduced into the forbidden gap by dopants used to make p–n junctions. Choose a few materials to demonstrate the trend. Make a similar study for excitons. At what temperature would you expect the donor levels due to (a) Se in GaAs, and (b) Si in GaAs to be fully ionized?

6
Physics of semiconductor interfaces

6.1. Surface of a semi-infinite semiconductor

Let us consider how the physical properties of a solid change near the surface. So far, we have ignored the fact that crystals are 'finite' objects. Indeed, we have based our models of the electronic structure on the assumption that we can view the crystalline lattice as an infinite regular repetition of atomic constituents in space. These atoms are characterized by microscopic potentials. Since the atoms in a solid sit close to each other, the microscopic atomic potentials overlap and give rise to the periodic crystal potential invoked in Chapters 2 and 3.

In the free-electron model of electronic structure, it is assumed that this overlap is so strong that the effective potential seen by electrons is nearly constant, so that the electrons may be regarded as 'free' of any interaction with the lattice. The energy spectrum of such a system is a parabola ($E_k \sim \hbar^2 k^2/2m$) and, at low temperatures, the states with energy $E_k < E_F$ are occupied whereas the states above E_F (the Fermi energy) are empty. We have become accustomed to drawing the electron energy with respect to the bottom of the valence states at $k = 0$. However, we can also think of this energy on an absolute scale, i.e. relative to the vacuum (Fig. 6.1a). According to this figure, the minimum energy required to remove an electron from the top of the reservoir at the Fermi energy (that is to take an electron from the metal to a point at a large distance from the metal) is $W = -E_F$.

This figure also indicates that, in order to account for the existence of the surface, we must split the crystal into two semi-infinite parts; as a result the potential near the surface contains a tail that decreases slowly to the zero energy value in the direction towards the surface and into empty space. We have destroyed the translational symmetry of the crystal potential near the surface. If the crystal potential is changed, it follows that the distribution of the valence electrons near the surface is also altered from its bulk value. This is illustrated schematically in Fig. 6.2. The broken line in this figure shows what the charge distribution—plotted in the direction perpendicular to the surface—would be if we ignored the existence of the surface. The shaded areas indicate that a certain amount of charge is transferred from the metal to the space just outside the surface. The spilled-out charge leaves behind a positively charged region in the crystal, whereas the outside region is negatively charged. As our common-sense model suggests, there is a double charged layer at the crystal surface, so that if we want to remove an electron from the crystal

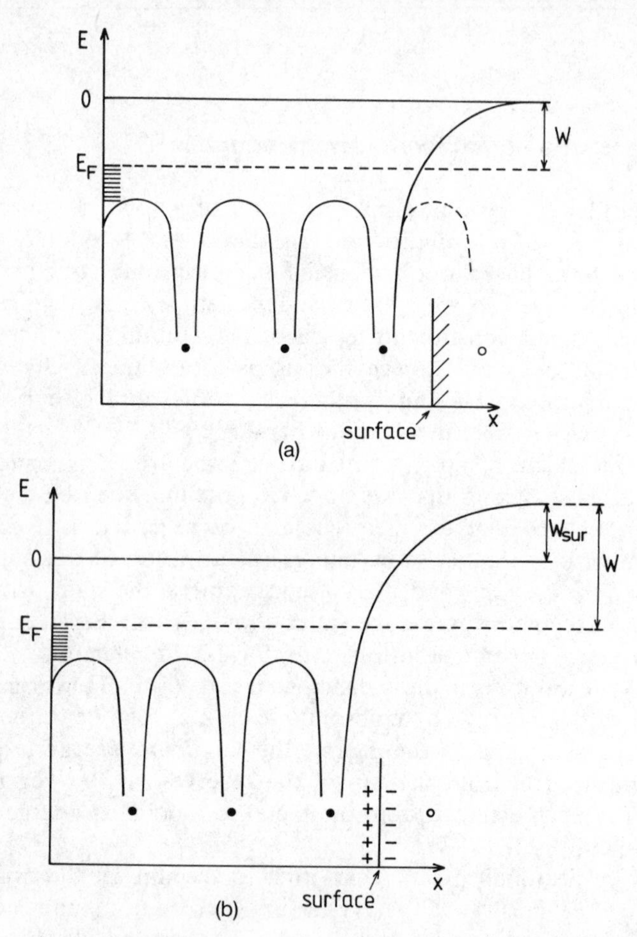

Fig. 6.1. (a) A sketch showing the position of the Fermi energy E_F relative to the vacuum ($E = 0$). The crystal potential is altered at the surface and the work function W is the energy needed to remove the uppermost electron from the crystal to infinity. The solid circles indicate the position of atoms in the lattice. (b) The work function W corrected for the contribution of the double charged layer W_{sur}. (c) Work functions W_A and W_B and Fermi energies $E_F(A)$ and $E_F(B)$ of two isolated metals A and B. (d) When A and B are in contact, the Fermi energies are level; ϕ is the contact potential.

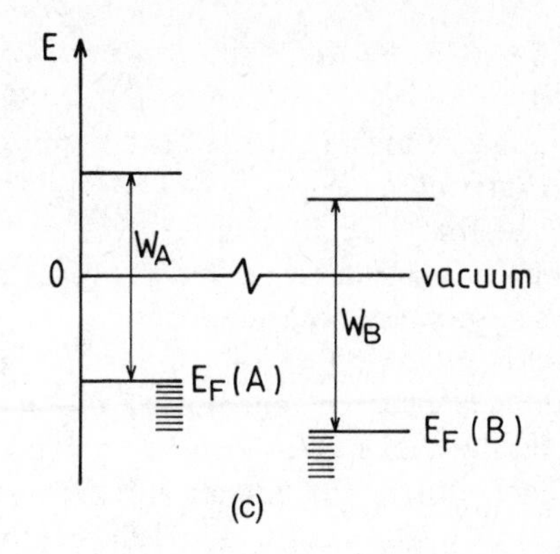

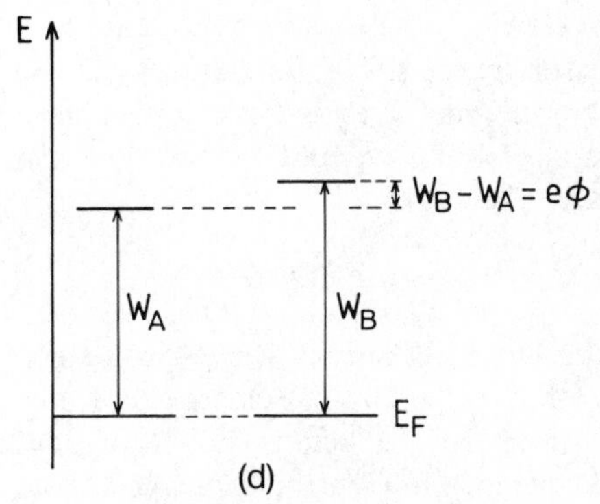

Fig. 6.1 (*continued*)

we must also supply an additional energy W_{sur} to overcome the double-layer potential. Hence the total energy needed is $W = -E_F + W_{sur}$ (Fig. 6.1b). This energy is often referred as the *work function*.

If we take two metals characterized by work functions W_A and W_B (Fig. 6.1c), and connect them by a conducting wire, we can measure the

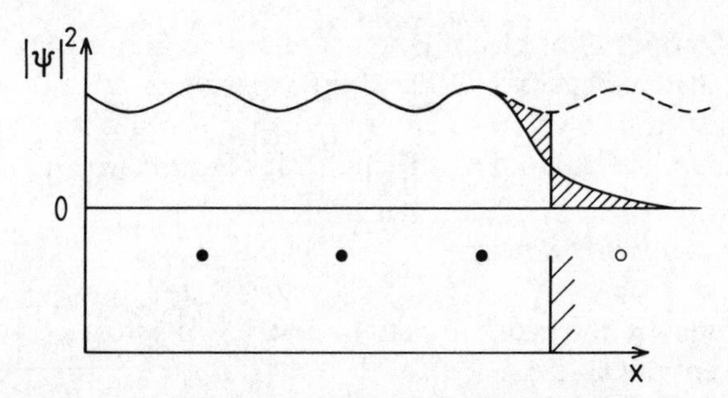

Fig. 6.2. The change in the electron charge density at the surface referred to in the text (hatched area); x is perpendicular to the surface.

potential energy difference between them, i.e. $W_B - W_A = e\phi$, the so-called contact potential (Fig. 6.1d). We know that when the two metals are connected, charge must flow until the Fermi energies are level. This is because the Fermi energy depends only on the electron density in the crystal. We can see that this procedure gives us the line-up of the electronic structure of the two materials and provides us with a concept that will be useful to us on many occasions later.

We can view the work function W of a given metal as a barrier an electron must overcome in order to get out of the crystal and into the ambient vacuum or dilute gas. Let us heat the metal and observe the current of electrons excited over the barrier and escaping into the region above the metal surface. This process is called thermionic emission. If we change temperature slowly enough as we take readings of the value of the electron current versus temperature we can assume that there is thermal equilibrium between the metal and the electron gas. In that case the probability that an electron is excited by the heat energy to a state of energy $\hbar^2k^2/2m$ above the barrier W is

$$f = \exp[-(\hbar^2k^2/2m + W)/k_BT], \tag{6.1}$$

where $\hbar^2k^2/2m$ is the kinetic energy of the excited electron. This accounts for the possibility that an escaping electron receives more than the minimum amount of energy, which is W. It means that the electron momentum k is zero when the electron is at the top of the barrier at W. Since W is typically about 1–2 eV, which is much larger than k_BT at normal temperatures, the application of the Boltzmann law in eqn (6.1) is well justified. The electron current flowing away from the surface is obtained by adding the contributions from electrons with velocities $v_x = \hbar k_x/m$, where x is the direction perpendicular to the surface. The current density j is

$$j = -e\int_{k_x>0} v_x f\rho \, \mathrm{d}k, \tag{6.2}$$

where ρ is the density of electron states of momentum k. We know from Chapter 1 that this density is $2/(2\pi)^3$, i.e. that there are two electrons of the same value of k in volume $(2\pi)^3$. We can substitute for f and ρ in eqn (6.2) and obtain the well-known Richardson–Dushman equation:

$$j = -\frac{em}{2\pi^2\hbar^3}(k_BT)^2 \exp[-W/k_BT]. \tag{6.3}$$

The magnitude of the work function (barrier W) for a few important metals is given in Table 6.1.

In setting up a simple model of the surface of a semi-infinite metal, we have idealized our picture quite considerably by assuming that we can

Table 6.1. The observed values of work function W for some metals

Metal	W (eV)	Metal	W (eV)
Na	2.35	Al	4.25
K	2.22	In	3.8
Cs	1.81	Ga	3.96
Cu	4.4	Sn	4.38
Ag	4.3	Pb	4.0
Au	4.3	Sb	4.08

split a crystal without altering the position of atoms in the crystalline lattice near the surface. The electron relaxation we did consider was assumed to be independent of the geometrical details of the surface. We also tacitly assumed that there are no imperfections at the surface. When we wanted to assess the contact potential, i.e. the potential barrier between two different metals, we ignored any changes that might occur as a result of interactions between adjacent planes of the different species of atoms forming the interface. In practice, these approximations are too crude to model real surfaces, and their breakdown is associated with the existence of new 'surface' states that significantly affect observable phenomena characteristic of surface physics. However, the idealized picture of a crystal surface is a useful one when we come to consider the interface between two solids. In particular, when the interface is formed by materials of similar crystalline structure, the physics of this interface can be viewed in terms of a superposition of two macroscopically well-defined materials whose bulk properties are preserved. It is precisely such interfaces that constitute the backbone of most useful semiconductor microstructures, and that is why the idea of an abrupt ideal interface introduced in this section will be very helpful. Structures dominated by features peculiar to rough reconstructed surfaces or interfaces play no part in the subtle art of engineering novel materials and device concepts in low-dimensional systems and will not be considered here.

6.2. Metal–semiconductor interface

Let us now consider what happens when a metal is deposited onto a semiconductor. As in the case of metal–metal contact, we shall assume that any unwanted effects at the interface, such as geometrical irregularities, can be ignored and regard the interface as an abrupt planar contact between two ideal crystalline species. The metal–semiconductor

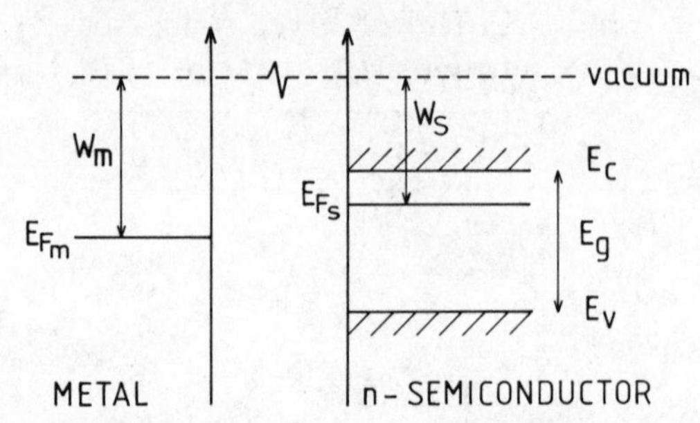

Fig. 6.3. The band diagram of an isolated metal and an n-type semiconductor with their work functions and Fermi energies relative to the vacuum indicated. Since the semiconductor is doped with donors, its Fermi energy lies closer to the conduction band edge E_c.

contact is one of the most common structures used in electronics. We shall describe the essence of the physics peculiar to this type of structure along the lines of the argument developed for p–n junctions and for ideal surfaces.

In Fig. 6.3 we have a shorthand picture of the electronic structure of a simple metal and, on the right-hand side, that of an n-type semiconductor. The position of the free-electron Fermi energy is given relative to the vacuum for both materials. In this figure, these two materials are viewed as independent species, i.e., disconnected. When the two materials are joined together, charge must flow until an equilibrium is established. In the case in question, electrons flow from the semiconductor into the lower-lying (empty) states available in the conduction band of the metal. As a result of this charge transfer, there is a region in the semiconductor where there is shortage of electrons and which is therefore positively charged. On the metal side of the interface, there are more electrons than there are in a neutral metal. This means that we have a double charged layer that gives rise to an electrostatic potential (sloping on the semiconductor side of the interface) that acts against the flow of electrons. As this space charge region grows, the corresponding potential barrier increases until it eventually stops electrons from spilling further into the metal.

It is customary to add the barrier potential due to the double layer to the 'ideal' band diagram shown in Fig. 6.3. This leads to the diagram in Fig. 6.4, illustrating the resulting band bending, i.e. the gradient in space (in the direction perpendicular to the interface) of the apparent band edges. Figure 6.4 represents the equilibrium result characterized by the depletion layer of width d analogous to that in a p–n junction. On the metal side, the potential gradient is negligible. This is because the electron density in a conductor (metal) is very high (typically or order

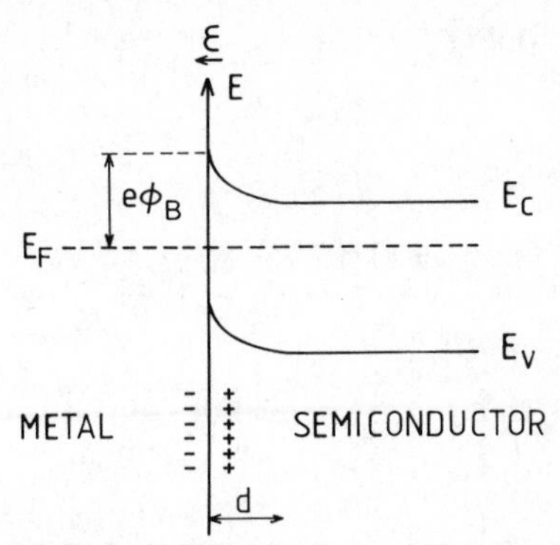

Fig. 6.4. The equilibrium band diagram of a metal–semiconductor interface formed by the materials shown in Fig. 6.3: ϕ_B is the contact potential; d is the width of the depletion layer. The direction of the electric field across the interface is also indicated.

$10^{24}\,\text{cm}^{-3}$). Consequently, in a metal electrons can move easily from region of higher electron density to lower and any potential gradient is instantly wiped out. Another way of putting it is to say that the dielectric constant of a metal is for practical purposes infinitely high. In a doped semiconductor, the free carrier concentration is of order $10^{19}\,\text{cm}^{-3}$ or less, since it is given by the concentration of dopants (donor and acceptors) and the metallic effect can be neglected. A typical value of the dielectric constant of an intrinsic (pure) semiconductor is of order 10 (see Table 4.1). This, too, reduces the magnitude of the space charge potential as well as the width of the depletion layer in the semiconductor, but the layer width is by no means negligible.

We can estimate the width of the depletion layer d in the manner outlined in Chapter 4 in connection with p–n junctions. From Gauss's theorem the electric field $\mathscr{E} = 0$ at $x = d$; x is perpendicular to the interface (Fig. 6.5). If the donor concentration is N_d and the semiconductor dielectric constant is ϵ, we have for the electric field

$$\mathscr{E} = eN_d \frac{(x-d)}{\epsilon_0 \epsilon} = -\frac{\mathrm{d}V}{\mathrm{d}x}. \tag{6.4}$$

After integration we obtain

$$V(x) = -eN_d \frac{(x^2 - 2dx)}{2\epsilon_0 \epsilon} \tag{6.5}$$

and

$$V(d) = \phi = \frac{eN_d d^2}{2\epsilon_0 \epsilon}, \tag{6.6}$$

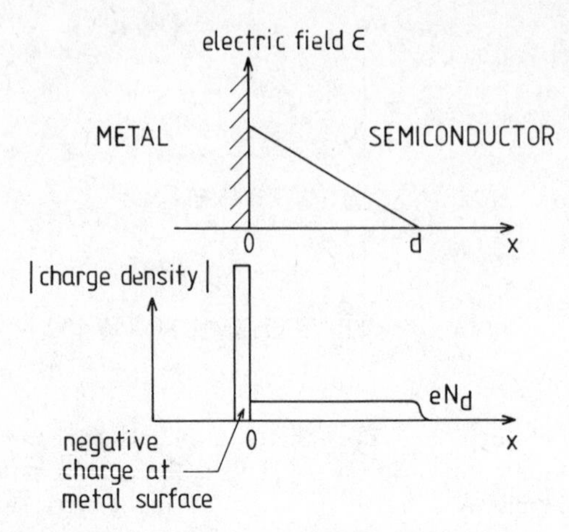

Fig. 6.5. The electric field $\mathscr{E}$ and the extrinsic charge density distribution at the metal–semiconductor interface. N_d is the concentration of (ionized) donors.

where $e\phi$ is the potential energy difference. If there is an applied (external) electric field, we add its magnitude to $e\phi$ in the same way as in the case of p–n junctions.

In Fig. 6.5 we can appreciate the asymmetric character of the metal–semiconductor contact potential. This asymmetry is an important factor in practical considerations, since in the forward bias (when the barrier is lowered by an external electric field) the electrons crossing the potential barrier have energies much larger than the Fermi energy in the metal. This is to say that they are injected with a large kinetic energy and we express that by calling them 'hot'. In fact, in this case electrons are the majority carriers. This is the opposite situation to that we encountered in p–n junction structures, where it is the minority carriers that are the main contributors to the forward current. The concentration of majority carriers is large and they can redistribute back much faster (typically 10^{-12} s), so that the switching time of a metal-semiconductor diode (the so-called Schottky barrier) is limited only by the charging time of the space charge capacitance evaluated in Chapter 4. The switching time of a p–n junction is much slower because the minority carrier lifetime is long (10^{-6} s). Finally, given the hot condition of the carriers, and the strong concentration gradients that can develop in the active part of this structure in the space charge region, the current–voltage characteristics must be corrected to account for contributions due to thermionic emission, discussed in Section 6.1, and electron and hole diffusion (Chapter 4). In both cases the magnitude of the potential barrier is of key importance.

If the doping concentration in the semiconductor is low ($\lesssim 10^{15}$ cm^{-3}), the band bending effect and its contribution to the potential barrier $e\phi$ is small, particularly in materials with large dielectric constant. If we neglect doping, we can determine the Schottky barrier height in the same way as we did above the contact potential in the case of two metals: we shall assume that some transfer of charge will occur, so that the Fermi level of the metal and that of the semiconductor lie at the same energy. The potential due to this charge transfer must be confined to a narrow region near the interface, and since it is inversely proportional to the dielectric constant of the semiconductor, and involves only a small amount of charge, we can as a first approximation neglect it. The barrier height is then given by the difference in energy of the semiconductor Fermi level E_{F_S} and the bottom of the semiconductor conduction band E_c (Fig. 6.6). Both these energies are intrinsic bulk parameters of the semiconductor in question, so that our theory predicts that the Schottky barrier height is independent of the metal species, i.e. in a given semiconductor it is the same for all metals. This remarkable and simple prediction has been well supported by experimental data. However, the model developed here is only qualitative and an accurate value of the barrier height must be obtained from a sophisticated calculation or from experiment. Also, there are metals that have a vigorous chemical reaction with the semiconductor surface and in such cases our model is not useful at all. But if we consider, say, gallium, aluminium, or indium deposited on high-quality silicon, we indeed obtain approximately the same Schottky barrier height. The magnitude of the Schottky barrier for gold deposited on some important semiconductors is given in Table 6.2.

Owing to the existence of the potential barrier indicated in Fig. 6.4, the metal–semiconductor structure behaves as a rectifier. The width of the depletion layer that is required in device applications can be computed

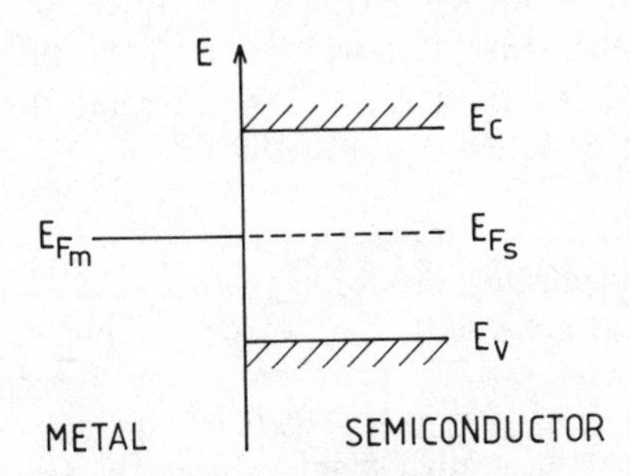

Fig. 6.6. The band diagram for an interface between a metal and an intrinsic semiconductor (the effect of charge transfer is ignored). The energy difference between the Fermi level at E_{F_s} and the bottom of the conduction band E_c is the Schottky barrier height.

Table 6.2. The fundamental gap E_g^f at room temperature, and the Schottky barrier height $e\phi$ for gold in contact with some semiconductors

	Si	Ge	InP	AlAs	GaAs	InAs	InSb
E_g^f (eV)	1.12	0.66	1.34	2.15	1.43	0.36	0.18
$e\phi$ (eV)	0.32	0.07	0.77	0.96	0.52	0.47	0.00

from eqn (6.6). In heavily doped semiconductors, the barrier width is only a few tens of angstroms. In Chapter 8 we shall evaluate the probability that a free carrier can 'tunnel' through such a barrier so that it can reach the semiconductor without having to gain enough energy to overcome the barrier height. In thin barriers, this ohmic 'tunnelling' current often dominates, particularly at low temperatures, over the non-ohmic contributions to current due to diffusion and thermionic emission. An increase in applied electric field can be used to enhance the tunnelling current and to ensure ohmic behaviour.

6.3. Metal–oxide–semiconductor structure

The electronic structure of a more complex system in which the metal is separated from the semiconductor by thin insulating layer (i.e. a layer of a material with very large forbidden gap) is shown in Fig. 6.7. The most common configuration involves a simple metal such as aluminium, a thin layer of SiO_2, which is a natural oxide of silicon, and p-type silicon. Hence, this structure is commonly referred to as the metal–oxide–semiconductor (MOS) structure. The band diagram in Fig. 6.7a is constructed as in the Schottky barrier case. Since the band edges of the oxide are several electron-volts away from those of the semiconductor (the band gap of SiO_2 is about 10 eV), their actual position is of no great importance and will be left out of our discussion. We shall again assume

Fig. 6.7. (a) The band diagram for a metal–oxide–semiconductor (MOS) structure. The semiconductor is p-type; u is the width of the oxide layer. (No charge transfer included.) (b) The equilibrium band diagram for the system in (a), with the band bending due to the double charged layer indicated. (c) The band diagram of (b) with a positive applied voltage V_a. (d) The band diagram of the MOS system in the strong inversion condition $\phi_s = 2\Delta$; l is the width of the depletion layer. In this diagram energy is measured from the top of the valence band. E_g is the fundamental gap of the semiconductor (about 1.17 eV for silicon at $T = 0$ K).

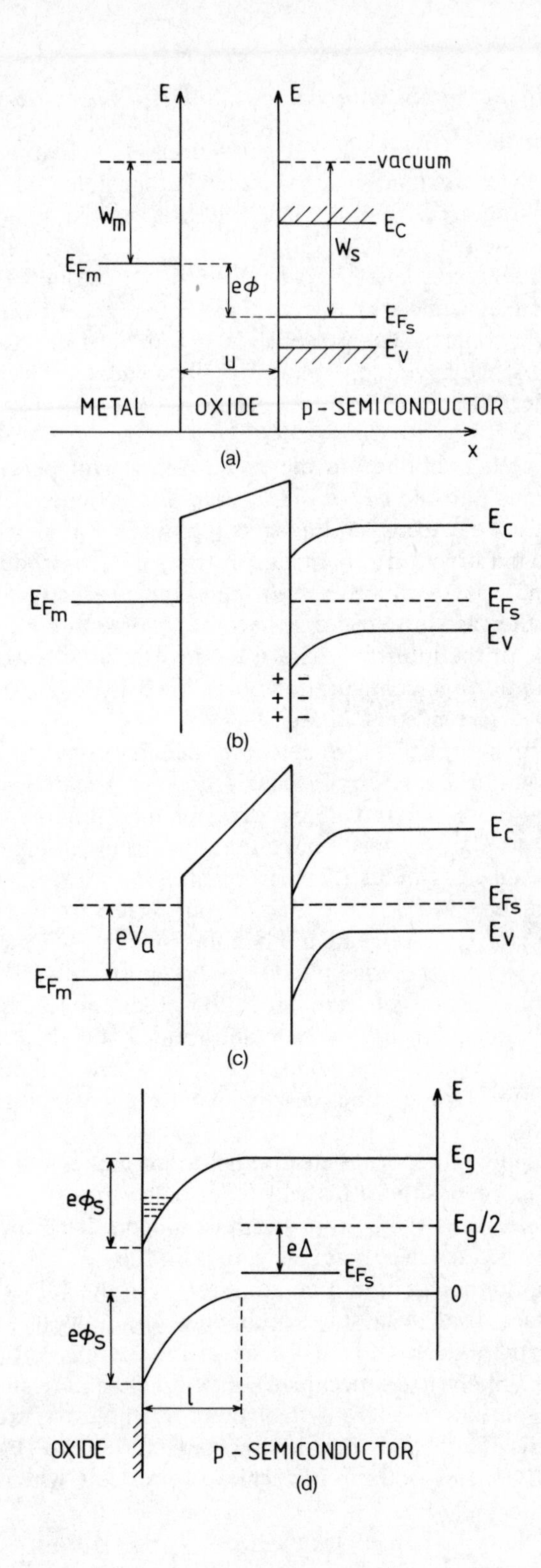
E
E
vacuum
W_m
E_C
W_s
E_{F_m}
$e\phi$
E_{F_s}
E_V
u
METAL
OXIDE
p - SEMICONDUCTOR
x
(a)
E_c
E_{F_m}
E_{F_s}
E_v
(b)
E_c
E_{F_s}
eV_a
E_v
E_{F_m}
(c)
E
E_g
$e\phi_s$
$E_g/2$
$e\Delta$
E_{F_s}
0
$e\phi_s$
l
OXIDE
p - SEMICONDUCTOR
(d)

that the interfaces are abrupt and geometrically and otherwise perfect enough that the approximations introduced earlier hold good. We can see that if we took the oxide (middle) layer out of Fig. 6.7a, we would end up with the metal–semiconductor diagram of Fig. 6.3 (the only difference being that here we have a p-type semiconductor). This diagram represents the electronic structure as if the layers were independent and not in electrical contact. The presence in Fig. 6.7a of an insulating oxide layer means that we have there what is really a capacitor, with one metal plate electrode (Al), and the p-type semiconductor serving as the other electrode.

A positive voltage applied to the metal relative to the semiconductor attracts electrons into the silicon-side of the SiO_2–Si interface. When the density of these electrons in the silicon interface region exceeds the density of the majority carriers in silicon (i.e., holes introduced there by doping the material with acceptors), the semiconductor is said to be 'inverted' (in the case in question it becomes n-type in a narrow layer on the silicon side of the interface). Since the inverted MOS structure is one of the most important configurations employed in microelectronics, we shall describe it here in some detail.

Let us return to Fig. 6.7a, which shows the electronic structure before the charge equilibrium is established. Here the Fermi energies of the metal and p-Si are separated by the potential energy step of $e\phi$. Although SiO_2 is an insulator, the oxide layer is thin enough for electrons to tunnel through it without the help of an external field until the two Fermi energies are leveled. The charge transferred from the metal into the semiconductor gives rise to band bending and a space charge region is established on the semiconductor side. (No potential gradient can remain at the metal interface). The gradient in the potential across the oxide is also indicated schematically in the equilibrium band diagram shown in Fig. 6.7b. This gradient represents a built-in electric field due to the difference in the work functions of the metal and semiconductor illustrated in Fig. 6.7a.

If we now apply an external electric field that makes the metal side of the structure more positive, this will repel holes from the interface and attract more electrons there from the bulk silicon. This means that the band bending is increased, as indicated in Fig. 6.7c. The electrons attracted from the bulk can find no empty slots in the valence band close to the interface and will start occupying the lowest states in the conduction band (which is bent downwards near the interface and is therefore more likely to be occupied). Eventually, there must be in the interfacial region more electrons than there are holes. If we neglect the built-in electric field and the small contribution to band bending associated with it, this inversion is achieved precisely when the external electric field exceeds the critical threshold value at which the field-induced band bending $e\phi_s$ equals twice the separation of the p-type semiconductor Fermi energy and the midgap energy (Fig. 6.7d), i.e., $\phi_s = 2\Delta$. This is the condition that defines the magnitude of the external electric field at which the bulk density of holes equals the density of electrons in the interface region. We can calculate these carrier densities in the manner described in Chapter 4. The concentration of holes at temperature T is $(N_a \gg n_i)$

$$p(\cong N_a) = n_i \exp(e\Delta/k_B T), \tag{6.7}$$

where N_a is the concentration of acceptors and n_i is the intrinsic carrier density in silicon. This equation also states that for $\Delta = 0$ we have an intrinsic materials with the Fermi energy at the midgap position and with p equal to the intrinsic carrier concentration.

In the absence of band bending, the concentration of electrons is, from eqn (6.7),

$$n = n_i \exp(-e\Delta/k_B T), \tag{6.8}$$

since $pn = \text{const}$. If the bias potential is $\phi_s = 2\Delta$ as shown in Fig. 6.7d then the electron concentration must be

$$n = n_i \exp[(-e\Delta + 2e\Delta)/k_B T] = p. \tag{6.9}$$

We can estimate the width of the depletion region l using the depletion approximation leading to eqn (4.25). The total charge of ionized acceptors per unit area localized in the region of width l is $-eN_a l$. The barrier potential is $\phi = eN_a l^2/2\epsilon_0\epsilon$, where $\phi = 2\Delta = (2k_B T/e)\ln(N_a/n_i)$. Hence we obtain

$$l = 2\left(\epsilon_0\epsilon\frac{k_B T}{e}\ln\left(\frac{N_a}{n_i}\right)\Big/ eN_a\right)^{1/2}. \tag{6.10}$$

For example, if $N_a = 2\times10^{16}\,\text{cm}^{-3}$ we get $l = 0.2\,\mu\text{m}$ at 300 K; l is indicated in Fig. 6.7d. Note that any additional increase in the positive bias can only increase the number of electrons available at the interface region but does not change l.

So far, we have disregarded the finite capacitance C_{ox} of the oxide and the correction due to the difference in the work functions (Fig. 6.7a,b). Let us take the oxide capacitance first. The total charge of ionized acceptors per unit area in the space charge region l is

$$n = -eN_a l. \tag{6.11}$$

The potential drop due to n is $V_s = -eN_a l/C_{\text{ox}}$. (The capacitance is approximately $C_{\text{ox}} = \epsilon_0\epsilon_{\text{ox}}/u$, where u is the thickness of the oxide layer; $\epsilon_{\text{ox}} \simeq 3.9$.) Hence, the threshold voltage V_T required to achieve inversion is obtained from

$$V_T - \frac{eN_a l}{C_{ox}} = 2\Delta. \tag{6.12}$$

The potential drop due to the work function difference is simply $e\phi = W_s - W_m$, as indicated in Fig. 6.7a. Thus the final expression for V_T is

$$V_T = 2\Delta + \frac{eN_a l}{C_{ox}} + \phi. \tag{6.13}$$

The condition (6.9), which serves as a basis for the more accurate expression for V_T in eqn (6.13), is often referred as the strong inversion condition.

6.4. Electron confinement in inversion layers

The width of the space charge region l indicated in Fig. 6.7d also marks the region over which band bending, associated with the charge transfer (the double charged layer) at the interface between SiO_2 and Si, is significant. This band bending is nothing less than the spatial variation of the average electrostatic potential of the kind considered in our earlier discussion of p–n junctions and Schottky barriers. We have tacitly assumed that this spatial variation is slow and smooth so that the electronic structure and the key bulk physical parameters such as the band gaps, band widths, effective masses, etc., remain unchanged. In short, we set up these diagrams on the assumption that the electrostatic potential representing the band bending can be superimposed upon the bulk crystal potential that we considered in Chapters 2 and 3 and that gives rise to the band structure of an infinite perfect crystal. We describe electrons and holes residing in quantum states near the conduction and valence band edges as 'free' particles with energy of const. $+ \hbar^2 k^2/2m^*$, where m^* is the bulk effective mass associated with the band extremum in question and k is the wave vector measured from the position of this extremum (for example, the bottom of the conduction band at the centre of the Brillouin zone).

Let us consider such an electron attracted towards the interface from the bulk silicon crystal and let us assume that the MOS structure is in the strong inversion configuration of Fig. 6.7d. As our electron approaches the interface region, it will, in addition to the 'normal' bulk crystal potential accounted for by m^* and k, experience the electrostatic band bending potential. As we can see in Fig. 6.7d, this potential is more and more attractive to electrons as we get closer to the interface. Hence, it is as if this electron was caught in an attractive one-dimensional well whose depth is gradually increasing from zero at the bulk band edge to some maximum near the interface with SiO_2.

In order to find out in what way the electron structure changes (what additional states there might be) relative to the unperturbed bulk band structure, it is necessary to solve the Schrödinger equation of the form

$$\left\{-\frac{\hbar^2}{2m^*}\frac{d^2}{dx^2}+V(x)\right\}\psi(x)=E\psi(x), \tag{6.14}$$

where the energy E is measured from the conduction band edge of bulk silicon and $V(x)$ is the band bending potential energy shown in Fig. 6.7d; x is perpendicular to the interface. We could now substitute for $V(x)$ the explicit form obtained from the Gauss's law and solve eqn (6.14) numerically provided that the parameters required in the calculation of the function $V(x)$ are available. The effective mass is that of bulk silicon. A rough estimate can be made if we invoke the fact that $V(x)$ resembles a triangular well $(-eFx)$. Once the form of $V(x)$ is specified, we can solve eqn (6.14) approximately by choosing a trial wave function. For example, take

$$\psi(x)=\left(\frac{b^3}{2}\right)^{1/2} x\exp\left(-\frac{b}{2}x\right) \tag{6.15}$$

such that the trial wave function is properly normalized:

$$\int_0^\infty |\psi(x)|^2\,dx=1. \tag{6.16}$$

Since we assume that the conduction band edge of SiO_2 is high above the silicon conduction band, this represents an 'infinite' barrier for motion into the oxide. The boundary condition for eqn (6.14) is $V(0)=\infty$, so that the amplitude of the trial wave function must be zero at $x=0$ and for $x<0$, i.e. inside the oxide layer. Substitute into eqn (6.14), multiply by $\psi(x)$ and integrate the equation to obtain an expression for energy E:

$$E=\int_0^\infty \psi(x)\left\{-\frac{\hbar^2}{2m^*}\frac{d^2}{dx^2}\right\}\psi(x)\,dx+\int_0^\infty \psi(x)V(x)\psi(x)\,dx. \tag{6.17}$$

Since $V(x)$ is fixed, the only adjustable parameter is b of eqn (6.15), which measures the rate of decay of our trial wave function. The best solution is that which gives the lowest energy (remember that E is negative, since the zero energy is set to lie at the conduction band edge of the bulk silicon). This variational approach, which is based on the general principle that a physical system always acquires a minimum energy configuration, is quite common in quantum mechanics. Clearly,

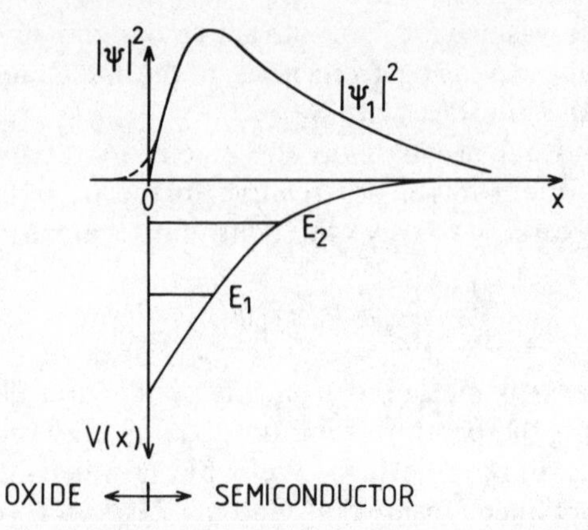

Fig. 6.8. The band bending potential $V(x)$ and the confined discrete energy levels E_1 and E_2, measured from the bottom of the conduction band of the semiconductor. The charge density $|\psi_1|^2$ distribution in the direction perpendicular to the interface (along x) associated with the wave function of the lowest (ground) state is also shown. The broken line indicates the penetration of charge into the oxide that is obtained when the infinite-barrier assumption is removed.

we might wish to try a better (two- or three-parameter) trial function, hoping to improve the accuracy of our solution. The better the choice of the wave function, the deeper the energy level.

The general form of the solution is sketched in Fig. 6.8, together with the potential well $V(x)$ and the energy levels one expects to obtain from eqn (6.17). Both E and V are measured from the bottom of the silicon conduction band.

The approximate wave function chosen as an example in eqn (6.15) is zero at the interface, and since we have assumed that the barrier is infinite there, we can expect zero penetration of the electron charge into the oxide. In reality, the barrier height is, of course, finite (about 1.5 eV) and some charge penetration takes place. In a more detailed calculation we would, therefore, expect to obtain a tail, indicated in Fig. 6.8 by the interrupted line. It transpires that only about 0.1% of the electron charge actually 'leaks' into the oxide. The wave function peaks well away (about 50 Å) from the interface and decays slowly in the perpendicular direction, into the semiconductor layer. We shall see that both these results are important for practical considerations.

Let us now turn to the electron energies we expect from eqn (6.17). As in the case of a one-dimensional rectangular square well (Appendix 2), the solutions are a discrete set of levels of energy E_1, E_2, etc. The number of these levels depends on the depth and range of $V(x)$. Since the number of electrons available near the interface is relatively small, only the lowest level will normally be populated. The charge density curve in Fig. 6.8 shows that in the direction perpendicular to the oxide–semiconductor interface, electrons in the ground state level are localized in a narrow layer close to, but not in significant contact with, the interface.

In the direction parallel to the interface, in the y, z plane, we expect the electronic structure of the semiconductor to be unchanged (bulk-like) i.e. $E_{\parallel} = \hbar^2 k_{\parallel}^2/2m^*$. Hence the total electron energy is

$$E_T = E_{\parallel} + E_{\perp} + E_c \tag{6.18}$$

where E_c is the energy of the conduction band edge of bulk silicon, and $E_{\perp}$ is the solution of the one-dimensional Schrödinger equation (6.17) for the perpendicular potential $V(x)$. Thus we can visualize the electron distribution near the interface as a very thin or quasi-two-dimensional sheet of electron gas. In the y, z plane, the electrons are free to move as in a bulk silicon crystal, whereas in the x direction they are confined near the interface. Since the electron energy has no dispersion in the x direction, the electron states fall into two-dimensional sub-bands, in the space of k_y and k_z, that can be labelled in terms of the confined discrete levels E_1, E_2, etc, shown in Fig. 6.8.

The change of dimensionality has profound effect upon the basic properties of the system. For example, as a result of confinement which introduces discrete levels in one direction, the density of states of this quasi-two-dimensional system is a step-like function (see problem 1.2), i.e. it is a constant for a given band and not a parabolic function of energy characteristic of a three-dimensional solution outlined in the first chapter. We shall return to the properties of confined states in thin quasi-two-dimensional layers in the next chapter and discuss the electronic structure of such systems in greater detail.

Let us summarize the predictions derived from the quantum-mechanical treatment of the band bending potential. If the external voltage applied to the MOS structure reaches the threshold value and the band bending potential is large enough, electrons flow towards the interface where they are confined in a narrow region. This is pictured schematically in Fig. 6.9 where we can also see the planar arrangement normally used in devices. If we apply an electric field along the interface, electrons will move as indicated by arrows in Fig. 6.9b. Since the amplitude of the electron wave function at the oxide–semiconductor interface is negligibly small, the electron motion will be insensitive to the properties of the interface and the oxide itself, and will depend only on the (bulk) properties of the silicon, and on the band bending potential. This means that the importance of any irregularities and imperfections at

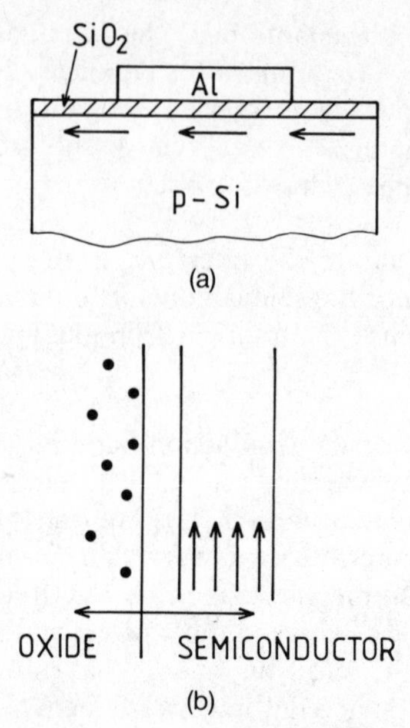

Fig. 6.9. (a) An MOS planar structure. The arrows indicate the flow of electrons in the inversion layer in the direction parallel to the oxide–semiconductor interface plane. (b) Perpendicular view of the electron channel in (a). We can see that the electrons are well separated from imperfections on the oxide side of the interface.

the interface is greatly diminished. Collisions with such imperfections would reduce the electron mobility.

We can now justify the assumptions made at the beginning of our account of the properties of this structure. The lattice spacing in SiO_2 and Si is very different. It follows that the mismatch at the SiO_2–Si interface is considerable and the interface is not likely to be an ideal, abrupt structure. However, in the light of the above discussion, this has little effect on confined electrons and the idealization is quite harmless.

The flow of electrons in the two-dimensional channel can be switched off simply by turning off the applied voltage that is responsible for the strong inversion. The electrons then simply return into the bulk silicon and the active channel near the interface shown in Fig. 6.9 disappears.

The form of the band bending potential, the degree of electron confinement, and the position of discrete levels described in the above paragraphs are quite general and depend only on the doping concentration and on bulk band structure parameters of the semiconductor. The prescription given here can therefore be used to optimize the width of the active channel and other properties of this type of structure by considering materials with other band structure parameters, dielectric constants, and doping. Such considerations are particularly relevant when we desire to reduce the size of MOS structures for application in submicron devices.

6.5. Heterojunctions

The structure that arises when two different semi-infinite semiconductor crystals are joined together is called in the semiconductor literature a '*heterojunction*'. As in our models of other interfaces, we shall assume that the crystal potential and consequently the electronic structure of each constituent semiconductor remains unchanged (bulk-like) right up to the interface plane, where it changes abruptly into bulk properties of the other semiconductor. In this case, the structures of interest are those in which the constituents are similar materials, or at least materials whose lattice structure and lattice constants differ very little. Fortunately, most technologically important semiconductors possess the diamond or zinc-blende symmetry (Chapter 5) and are therefore quite compatible provided the lattice constants are similar. The lattice constant is, therefore, the parameter to use as a measure of compatibility of semiconductor materials forming a heterojunction. In the light of these considerations, we can say that heterojunctions are structures for which the simplifications (the concept of abrupt interface) we introduced at the beginning of this chapter are likely to fit best.

In our account of metal–semiconductor interfaces, we did not invoke the details of the dispersion relations (electron energies versus wave vector) in any detail and were content with the simplest description of the electronic structure near the band edges (the fundamental gap). Our assumption about abrupt perfect interfaces implies that the band structures (dispersion E_k versus k) are also bulk-like for both constituents, i.e., on both sides of the interface. Hence, to obtain the electronic structure of the heterojunction system, we simply line up the bulk band structures. The most important item on our agenda must then be the question concerning the relative position of the constituent band structures. This will give the magnitude of the discontinuity (potential step at the interface) between the conduction (valence) band edges of the two semiconductors, the so-called conduction (valence) band offset.

In Section 6.2 we introduced a general prescription for lining up two free-electron-like crystals. We argued that the mean electron charge density must be the same on both sides of the interface. Consequently, the Fermi energy—which depends only on the electron density, and which in the nearly-free-electron model lies exactly in the middle of the vertical gap (Fig. 5.1) separating the occupied and empty states of an intrinsic semiconductor at zero temperature—must also represent the midgap energy level of both constituents. Small differences in electron density due to different lattice constants are removed by charge transfer across the interface. Since the electrostatic potential associated with such transfer is screened (divided) by the semiconductor dielectric constant, which is of order ten, and since the amount of charge transferred is expected to be insignificant, we shall assume that for intrinsic crystals the correction due to this double charged layer may be neglected. In other words, we start with a scheme analogous to that for a metal–semiconductor contact (Fig. 6.10a) and end up with a line-up shown in Fig. 6.10b. This is equivalent to a rigid shift of the two bulk band structures relative to each other so that the midgap positions are level.

In Chapter 5 we found a simple relationship between the dielectric constant ϵ^o and the (vertical) forbidden gap E_g obtained in the nearly-free-electron calculation developed in Chapter 2. The vertical (direct) band gaps for several semiconductors calculated from the empirical values of ϵ^o are listed in Table 5.1. If the two semiconductors forming the heterojunction have identical (or very similar) lattice constants, their

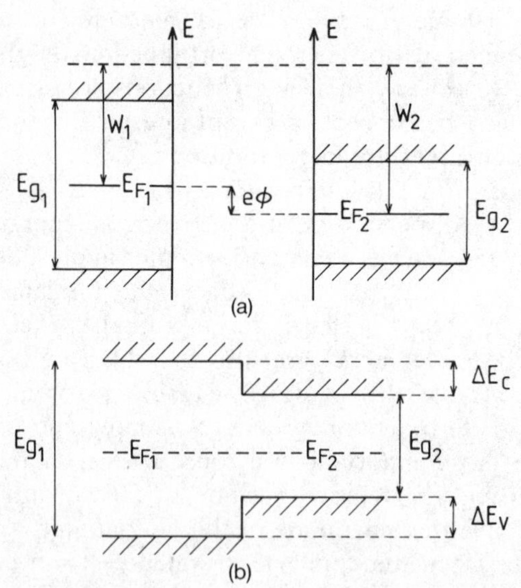

Fig. 6.10. (a) The band diagram of two different disconnected semiconductors 1 and 2, with energies shown relative to the vacuum. (b) The line-up of band structures of lattice-matched semiconductors 1 and 2 forming a heterojunction. ΔE_c and ΔE_v are the conduction and valence band offsets, respectively. It is assumed that the materials are undoped and that the effect of charge transfer at the interface is negligibly small. E_{g_1} and E_{g_2} are the fundamental gaps.

Fermi energies will coincide and all that remains to be done to obtain the valence band offset ΔE_v is to subtract the vertical average gaps as follows:

$$\Delta E_v = (E_{g_1} - E_{g_2})/2. \tag{6.19}$$

In Table 6.3 we list ΔE_v for several important heterojunctions and compare them with experimental values. Note that in the case of ΔE_v we seem to be remarkably successful in accounting for the observed trend. Recall that although we were able to use a similar model to predict the qualitative properties of the metal–semiconductor (Schottky) barrier, we had to be content with only an order of magnitude estimate of the barrier height. This is because in the present case the materials in contact are very much more similar, and, as we noted in Chapter 5, the uppermost valence band does fit the constant gap approximation in that it is very flat in all semiconductor crystals. If we attempt to use this model and eqn (6.19) to work out the conduction band offset, we would fail just as we did in the case of Schottky barrier heights. This is because the conduction band has a complicated structure with several minima whose relative

Table 6.3. The difference in the bulk lattice constant δa, the valence band offset ΔE_v calculated from the empirical gaps obtained in eqns (5.2) and (6.19); and the experimental value of the offset Exp, for some important heterojunctions[a]

Heterojunction	δa (Å)	ΔE_v (eV)	Exp(eV)
GaAs–Ge	0.00	0.47	0.56
GaP–Si	0.01	0.41	0.8
AlAs–GaAs	0.01	0.43	0.5
AlAs–Ge	0.01	0.89	0.95
ZnSe–Ge	0.00	1.51	1.52
ZnSe–GaAs	0.00	1.04	0.96
CdTe–α–Sn	0.01	1.22	1.0
CdTe–InSb	0.01	0.89	0.87
CdTe–HgTe	0.00	0.35	0.35
InAs–GaSb	0.05	0.20	0.46
Si–Ge	0.22	0.48	—
GaAs–InAs	0.40	0.37	0.17

[a] The material whose valence band lies lower is given first. The experimental data are expected to suffer from a substantial error because of the difficulties encountered in the measurements.

positions depend on details of the microscopic crystal potential and which in some cases lead to indirect gaps being smaller than the direct (vertical) ones as, for example, in silicon. Fortunately, once the valence band offset ΔE_v is known, the discontinuity (offset) in the conduction band ΔE_c is simply

$$\Delta E_c = |E^f_{g_1} - E^f_{g_2}| - |\Delta E_v|, \tag{6.20}$$

where $E^f_{g_1}$, $E^f_{g_2}$ are now the experimental values of the fundamental gaps of the two bulk constituents.

As we noted before, the fundamental gaps separating the uppermost occupied and the lowest empty states are not necessarily the same as the direct gaps computed from the nearly-free-electron model in Chapter 2 and invoked in Fig. 5.1 and eqn (6.19). If both constituents are direct gap materials, the prescription (6.20) is all we need to consider. If they are not, we must make sure that we determine the conduction band offsets by subtracting the energies of the conduction minima that are alike (i.e. associated with the same k point in the bulk band structure diagram). For example, we want the energy separation (discontinuity) between the lowest states at the Γ point, or that between the lowest states at the X point. We do not want a discontinuity between, say, the lowest points of the X and Γ valleys even if they represent the absolute conduction band minima in their respective crystals (which is the case, for instance, in a heterojunction formed by GaAs and AlAs). This is because, in our picture, the electron momentum (i.e. the wave vector k) cannot change at the interface. If we linked up an X valley with a Γ valley, the magnitude of the wave vector of an electron crossing the heterojunction would have to change.

To take a more specific example, if we consider a GaAs–AlAs heterojunction with $\Delta E_v = 0.5$ eV, an electron from the Γ valley of the conduction band of GaAs approaching the interface with AlAs (whose X valley lies below its Γ valley) experiences a barrier of height $E_\Gamma(\text{AlAs}) - E_\Gamma(\text{GaAs}) > 1$ eV, not $E_X(\text{AlAs}) - E_\Gamma(\text{GaAs}) \simeq 0.3$ eV, even though the latter is the energy difference between the lowest-lying bulk states in the conduction band of GaAs and AlAs.

To understand this 'rule' in some depth, we would have to venture well outside the scope of this course. However, one might get some additional insight into what we are doing by recalling the procedure employed in Appendix 2 in connection with the square well problem. There we start with electron wave functions (plane waves of momentum $\hbar k$) associated with energies lying at the bottom of the (free-electron) parabolic band of energies $\hbar^2 k^2/2m$. If the well has a finite depth, these waves are partially reflected from the wall, and form cosine- and sine-like standing waves inside the well with tails extending into the barrier. In order to find the form of the wave function satisfying the Schrödinger equation and exhibiting this tail, we have to match the wave functions (equate their radial derivatives) at the well boundary on both sides. The well boundary is just like the interface in the heterojunction problem. If the electron states on each side belong to a band with very different wave vector k and mass, they are quite unsuitable for this simple matching procedure. Only states with the same bulk k values can be matched to obtain the confining effect described in Appendix 2. We can say that the probability that a Γ electron approaching the interface 'sees' the X valley is very small.

By forming the discontinuity from the energy difference between the states of the same bulk momentum (wave vector), we are therefore retaining the picture in which the electron interaction at the interface can be modelled as a reflection from a step-like barirer. This very simple model nevertheless turns out to be quite realistic and will enable us to make a semi-quantitative estimate of such interactions in heterojunction microstructures. We also have a unique and simple prescription for finding both the valence and conduction band offset for any pair of semiconductors with not too dissimilar lattice constants, from readily available empirical (experimental) values of bulk band gaps and dielectric constants.

The line-up of the conduction bands of GaAs and $Ga_{0.5}Al_{0.5}As$ showing schematically both the step seen by an electron in the Γ valley of GaAs (solid line) and that by an electron in the X valley (broken line) is illustrated in Fig. 6.11. In Fig. 6.11a, we see the line-up corresponding to

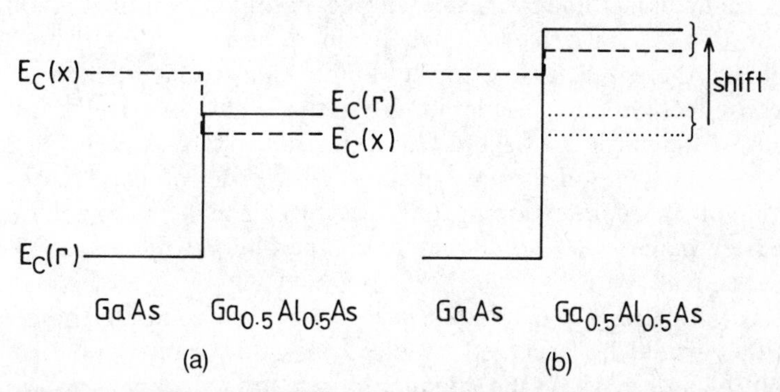

Fig. 6.11. The conduction band offsets for a GaAs–$Ga_{0.5}Al_{0.5}As$ heterojunction. The solid line shows the line-up of the Γ valleys and the interrupted line shows that of the X valleys. In (a) the valence band offset is so large that the top of the valence band of GaAs lies much higher (0.5 eV) than that of AlAs. In (b) the valence band offset is chosen to be only 0.15 eV, so that the order of the X minima (sign of the step potential seen by X electrons) is changed. The shift in the position of $E_c(X)$ and $E_c(\Gamma)$ due to the change in the valence band offset is indicated by the arrow, and the dots indicate the former position of these minima.

$\Delta E_v = 0.5\,\text{eV}$. In Fig. 6.11b the valence band offset is chosen to be different, namely $\Delta E_v = 0.15\,\text{eV}$. Clearly, having changed the magnitude of the valence band offset, we have had to shift rigidly the two band structures (shown in Chapter 5) relative to each other so that we have changed the height of the barrier seen by the electrons in the Γ valleys of GaAs. However, this has also changed the relative positions of other minima; for example, the sign of the step potential seen by electrons in the X valley is reversed.

Under normal circumstances, the X valley of a GaAs layer would be empty, since electrons first seek the states that have the lowest energy. We would have to supply an unphysically large concentration of carriers in order to fill the high-lying X-like levels. However, the secondary minima may play an important role in electron transport in microstructures, because electrons can be transferred by external fields into higher-lying levels.

So far we have assumed that the semiconductor materials forming a heterojunction are intrinsic. If there are dopants in one or both constituents, a space-charge region is established that leads to band bending. The spatial variation of the electrostatic potential associated with the existence of the double charged layer can be calculated in the same way as for the p–n junction, metal–semiconductor, and MOS structures. This potential is then superimposed upon the band diagram of the discontinuity (band offset), both at the conduction and valence band edges. We proceed along the same lines as we did in order to construct analogous diagrams in Sections 6.2–6.4. Since a heterojunction consists of two semiconductors, the space charge potential is non-zero on both sides of the interface, very much as in a p–n junction. The spatial variation of the potential due to the double charged layer, superimposed on the conduction band discontinuity (step) potential derived from band structures of intrinsic GaAs and $Ga_{0.7}Al_{0.3}As$ crystals, is sketched in Fig. 6.12. Since the band offsets are typically of the order of a few tenths of an electron-volt, moderate doping ($10^{17}\,\text{cm}^{-3}$) can make a contribution to the barrier height ($\sim 0.1\,\text{eV}$) that is comparable to the offset between intrinsic crystals.

The potential in Fig. 6.12 could also be used to compute the confined levels that would be occupied by electrons and localized on the GaAs side of the interface in the manner described in connection with the oxide–semiconductor case. The role of the insulating oxide is taken over by the $Ga_{0.7}Al_{0.3}As$. Since the heterojunction barrier separating the conduction band of GaAs and $Ga_{0.7}Al_{0.3}As$ is quite low compared to the barrier between Si and SiO_2, the wave function of the confined state will penetrate more strongly into the barrier material. It transpires that several per cent of the electron charge leaks into the barrier. Also, the

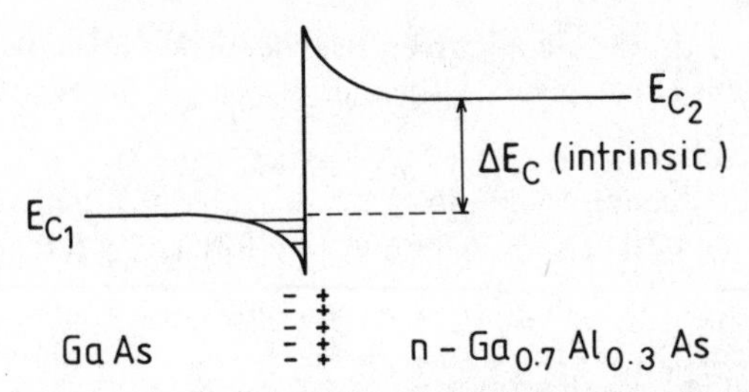

Fig. 6.12. The band diagram showing the bending of the conduction bands at a GaAs–$Ga_{0.7}Al_{0.3}As$ (n-type) heterojunction. ΔE_c is the conduction band offset for undoped structure. E_{c_1} and E_{c_2} are the conduction band (bulk) minima at the Γ point where the minimum of the fundamental gap of both these semiconductors occurs. The discrete confined levels are shown on the GaAs side of the heterojunction.

effective mass of the lowest conduction band valley of GaAs at the Γ point is very small (about 0.067 of the free-electron mass) compared to the effective mass at the lowest conduction band valley of silicon, which occurs at the X point. Hence the decay of the localized wave function representing confinement into the GaAs must be slower, and the width of the quasi-two-dimensional sheets of electron gas will be correspondingly larger. Such comparisons are of practical importance, since structures analogous to the silicon-based MOS system have been made of GaAs–$Ga_{1-x}Al_xAs$. We shall consider several device structures of this type later in this course.

Problems

6.1. Determine the barrier width at a metal–semiconductor interface assuming that the semiconductor is silicon and contains $10^{20}\,cm^{-3}$ donors. The barrier height from semiconductor to metal is 0.8 eV.

6.2. Find the current density for an Al–Si (n-type) Schottky barrier at 0.4 V 'reverse' bias (i.e. when the external field reduces the barrier height), at room temperature.

6.3. Determine the width of the depletion layer at room temperature for an MOS system with a p-type silicon containing an acceptor concentration of $1.5 \times 10^{17}\,cm^{-3}$.

6.4. Find the threshold voltage for strong inversion in a Si–MOS system at room temperature with an acceptor concentration of $2 \times 10^{16}\,cm^{-3}$. The oxide thickness is 0.1 μm and the metal–semiconductor barrier is -0.94 eV.

6.5. Find the value of the valence band offset such that (a) the conduction band offset of a GaAs–$Ga_{0.7}Al_{0.3}As$ heterojunction is exactly zero (ignore band bending effects) and (b) the step between the conduction band X minima is zero. Use the band structure data given in Chapter 5.

6.6. Find an ideal alloy of GaAs an InAs such that its lattice constant matches that of InP. Use the theory leading to results in Table 6.3 to predict the valence and conduction band offsets for a heterojunction formed by InP and the alloy.

6.7. Consider a heterojunction of undoped GaAs and n-type $Ga_{0.7}Al_{0.3}As$ ($N_d = 10^{17}$ donors cm^{-3}). Estimate the width of the space charge potential in GaAs on the condition that the electron population in the quasi-two-dimensional electron gas in GaAs equals the density of electrons depleted from the barrier material. Ignore the difference in the dielectric constant and assume that the magnitude of the contact potential equals to the magnitude of the conduction band offset. Draw a rough graph of the potential seen by an electron at the interface ($T \cong 300$ K).

6.8. In the structure of Problem 6.7, use a one-parameter trial wave function to find the rate of decay of the confined electron charge density in the lowest conduction miniband into the GaAs layer ($\int_0^\infty x^n \exp(-bx)\,dx = n!/b^{n+1}$).

6.9. Use the result of Problem 6.8 to determine the distance from the interface at which the probability of finding an electron has its maximum.

6.10. Draw a band diagram of a Si–GaP heterojunction using the theoretical value of the valence band offset given in Table 6.3 and the band structure data of Chapter 5. What is the magnitude of the conduction band offset?

6.11. Draw a band offset diagram for GaAs–$Ga_{1-x}Al_xAs$, $0 \leqslant x \leqslant 1$, using the experimental value of the valence band offset and the band structure data of Chapter 5 ($T \cong 4$ K).

7

Quantum wells, superlattices, quantum wires, and dots

7.1. GaAs–$Ga_{1-x}Al_xAs$ quantum wells

Our discussion in Chapter 6 of the band line-up at heterojunctions provides a useful starting point for considerations concerning semiconductor quantum wells. We have seen that a heterojunction formed by depositing a layer of, say $Ga_{0.7}Al_{0.3}As$ on a GaAs substrate leads to electronic structure that can be modelled as a step-like change (discontinuity) in the conduction and valence band edges. The height of this step, the so-called conduction or valence band offset, depends on the bulk band structures of the constituent materials and on their relative position. We developed a scheme, based on the nearly-free-electron model of Chapter 2, and extended further in Chapter 5, that enabled us to predict the band offsets from experimental values of the dielectric constant and bulk gaps. We found that for lattice-matched semiconductors whose fundamental gap is direct (i.e. the lowest point of the conduction band occurs at the same k point as the top of the valence band) our prescription was particularly simple. In other cases, the complexities of the bulk band structure (i.e. the multi-valley character of the conduction band) make the task of setting up heterojunction band line-ups more tricky. Nevertheless, we do have a clear rule for how to proceed in most cases of practical interest, and a semi-empirical conceptual picture based on the matching condition for the wave function at the interfaces to support our rule.

Most applications of semiconductor microstructures are based on heterojunctions of the simplest kind. By far the most frequently used materials are GaAs and $Ga_{1-x}Al_xAs$. The aluminium concentration is usually chosen to be around 30–35% to ensure that the potential barrier in the conduction band is large but that the alloy is still a direct gap material. It is these materials that we shall focus on here.

The condition concerning the abrupt nature of the interface is well satisfied in GaAs–AlAs structures and it is, therefore, possible to grow

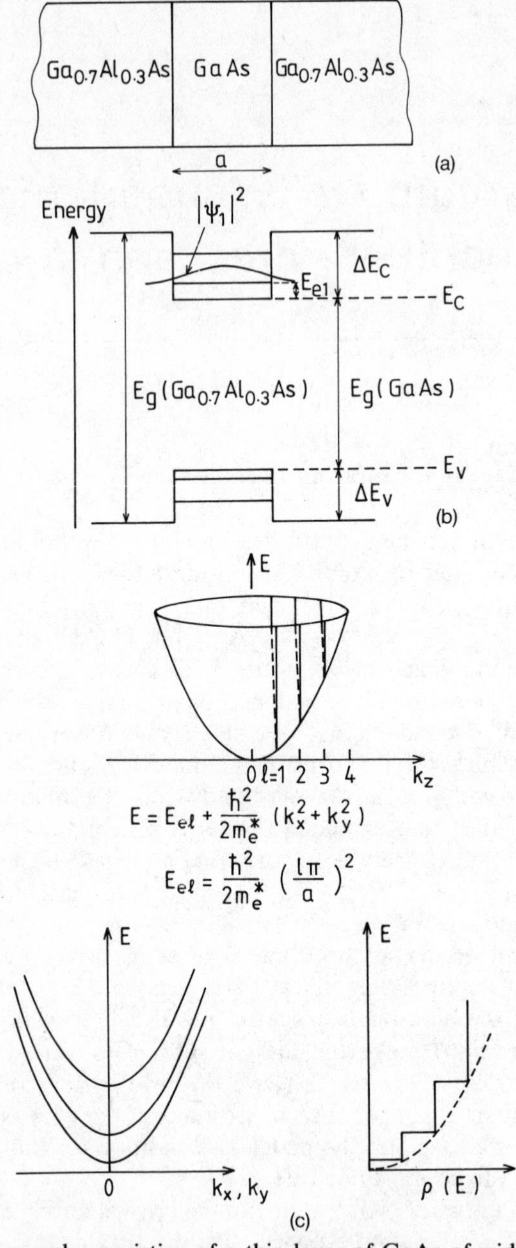

Fig. 7.1. (a) A sample consisting of a thin layer of GaAs of width a sandwiched between thick layers of $Ga_{0.7}Al_{0.3}As$. (b) The electronic structure of a GaAs quantum well. E_g(GaAs) is the forbidden gap of bulk GaAs. The lowest confined level in the conduction band lies E_{e1} above the conduction band edge of bulk GaAs E_c and its wave function is ψ_1. (c) The construction of the total energy of an electron in the conduction band of a quantum well structure; z is the direction perpendicular to the interface; ρ is the density of states (the broken line indicates the form of ρ for a bulk crystal).

more complicated systems consisting of several layers of such materials of well-controlled thickness. The simplest of these systems is shown in Fig. 7.1a. A thin layer of GaAs is sandwiched between two thick layers of the $Ga_{0.7}Al_{0.3}As$ alloy. We now have a double heterojunction system, which is often referred to as the '*quantum well*'. The reason for this label becomes clear when we look at the picture of the electronic structure shown in Fig. 7.1b. This picture is obtained simply by adding the results we described in Chapter 6 for a single heterojunction. Since the change of electronic structure is abrupt and the two materials are lattice matched, in the absence of doping (no band bending) the potential is just a superposition of two potential steps in the back-to-back position. The two heterojunctions are independent of each other and the band diagram remains the same irrespective of the well width (a) or the thickness of the alloy.

If one or both constituents are doped, then the band bending must be included in the same way as for a single heterojunction, and the potential profile changes accordingly. If the well is narrow, the band bending curves originating from each interface may overlap and the profile will depend on both the doping and the well width. However, as we noted in Chapter 6, the correction for doping is quite straightforward and in the following paragraphs we shall assume that the materials are intrinsic.

Let us consider how the existence of two heterojunctions affects the electronic structure. If a free electron is injected into the quantum well structure in Fig. 7.1, it will eventually find the region of lowest energy which, according to Fig. 7.1b, occurs in the GaAs layer. We shall assume that the GaAs layer is thick enough ($a > 20$ Å) that the layer has the properties characteristic of macroscopic GaAs crystals and thin enough ($a < 400$ Å) to ensure that a is shorter than the mean free path, i.e. that an electron is likely to traverse the well without undergoing a collision of any kind. If these two conditions are satisfied, we can, to start with, approximate the electron energy as $E_c + (\hbar k)^2/2m_e^*$ where m_e^* is the bulk effective mass of the lowest conduction band of GaAs and k is zero at the band edge (at E_c). The second condition allows us to think that our electron in the well can 'see' both barriers defining the well. The electron then makes reflections off the barriers (whose height is ΔE_c). In this way we have arrived at a situation that is, from a mathematical point of view, entirely analogous to that considered in Appendix 2, where we outline the solution to the 'particle in a box' problem in one dimension. The only difference is that the electron mass is m_e^*. This accounts for the fact that our electron is placed at the bottom of the conduction band of GaAs and not in free space. The magnitude of the effective mass reflects the strength of the crystal potential and, as we have seen in Chapter 5, varies distinctly from material to material. However, the band structures of the two semiconductors forming our quantum well structure are very similar near the band edges and we shall therefore ignore the difference in effective masses between GaAs and $Ga_{0.7}Al_{0.3}As$. As a result we can view the problem of confinement in a quantum well of Fig. 7.1 literally in terms of the simple particle in a box picture.

The effect of confinement upon the electron energy is to turn the energy spectrum in the direction perpendicular to the interface into a series of discrete levels. Two such levels are indicated in the well shown in Fig. 7.1b. Since these levels owe their existence to the effect of reflection at the interfaces, we must expect the electron wave function inside the well to have the form of a standing wave, i.e. $\psi \rightarrow \sin(\pi l z/a)$, where a is the well width, z is the axis perpendicular to the interface, and $l = 1, 2, \ldots$. Outside the GaAs well, the wave function ψ decays exponentially ($\psi \rightarrow \exp(-k_2 z)$, where $k_2 = (2m_e^* E/\hbar^2)^{1/2}$. E is measured from the bottom of the conduction band of GaAs at E_c. The ground state wave function with $l = 1$ is nodeless and the higher levels—the so-called excited states with $l = 2, 3, \ldots$, have 1, 2, etc. nodes. The smaller the effective mass, the further is the lowest confined level pushed from the conduction band edge E_c, and the more is the wave function tail expected to 'leak' out of the well. The lowest state ($l = 1$) is the best-confined of all levels (i.e. the least leaky one), and as l increases and the levels move towards to top of the well, their wave functions become more extended. The remarkable feature of this problem is that no matter how small the barrier height and the effective mass may be, and whatever the well width, there is always at least one confined state. It is possible to prove that this is only true for one- and two-dimensional potential wells.

So far we have considered the effect of confinement upon the electron energy spectrum in the z direction i.e. due to the reflections in the direction perpendicular to the interface. In the direction parallel to the interface, we expect the motion to be unaffected by the existence of the quantum well and the dispersion relations (energy versus wave vector) to be the same as in a macroscopic crystal. Hence, the total energy of an electron in the lowest confined level ($l = 1$) at the conduction band measured from E_c is $E_{e1} + (\hbar k_{\parallel})^2/2m_e^*$, where E_{e1} is the ground state solution of the Schrödinger equation with the one-dimensional potential (barrier height ΔE_c) indicated in Fig. 7.1b; $k_{\parallel}$ is the wave vector lying in the plane of the interface given by the axes x and y and measured from the conduction band edge of bulk GaAs ($k_{\parallel} = 0$ at E_c).

As in the case of confinement in silicon inversion layers considered in Chapter 6, when we plot this energy as a function of wave vector along x, y, we end up with 'minibands' (or 'sub-bands') of allowed energies separated by 'minigaps'. These sub-bands replace the continuous bulk band and can be labelled by the quantum number l we used to label the confined levels. This three-dimensional picture of the band structure at the bottom of the conduction band is shown in Fig. 7.1c.

In the upper part of Fig. 7.1c, we see the electron energy measured from the bottom of the conduction band of GaAs. The energy of the sub-bands as a function of the wave vector form parabolas with the lowest points at E_{el} where $l = 1$, 2, and 3 sub-bands are shown explicitly; 0 marks the point where the bulk conduction band of GaAs originates. If we turn to the lower part of Fig. 7.1c, on the left we can see the minibands that were also shown in the upper figure, but in this case drawn with the wave vectors parallel to the interface plane lying in the frontal position. (There is no dispersion in the direction z perpendicular to the interface. In that direction the confined states behave as bound states.) If we turn to the right-hand side of the bottom half of Fig. 7.1c, we see the density of electron states, ρ, associated with the sub-band structure on the left as a function of energy E. In a bulk (three-dimensional) parabolic band, the density of states goes as $E^{1/2}$, as indicated by the broken line. Owing to the effect of confienment, we have lost the dispersion in the z direction, so that the density of states is that of a two-dimensional continuum represented by a series of two-dimensional parabolic bands. If we apply the procedure for the derivation of the functional form of ρ, given in Chapter 1 for the three-dimensional case, to our two-dimensional bands, we obtain a series of steps (solid line), each originating from the onset of a sub-band. This means that the density of states at the bottom of the conduction band of our quantum well system is finite, whereas that for bulk GaAs is zero.

The procedure we have used to describe confinement of electrons at the conduction band edge can be also applied to the valence band. There we again have a step-like potential that will give us a well analogous to that in the conduction band. The height of the confining barrier is the valence band offset, which is shown in Fig. 7.1b as ΔE_v. We expect to obtain confined levels whose energy lies below the top of the valence band of bulk GaAs. The position of these levels relative to the bulk band edge is not the same as that in the conduction band, because both the barrier height (band offset) and the effective mass are in principle (and in practice) different quantities. It is customary to measure the energy of the confined levels at the valence band from the top of the valence band of GaAs. Then the picture of the total energy of the valence sub-bands is formally the same as that shown in Fig. 7.1c for the conduction sub-bands. However, as we learned in Chapter 5, in all semiconductors of interest to us there are two different bands at the top of the valence band, one that is rather flat (the 'heavy hole' band) and therefore has a large effective mass m^*_{hh}, and the other whose curvature is large and which has a small effective mass m^*_{lh} (the 'light hole' band). Hence we expect two

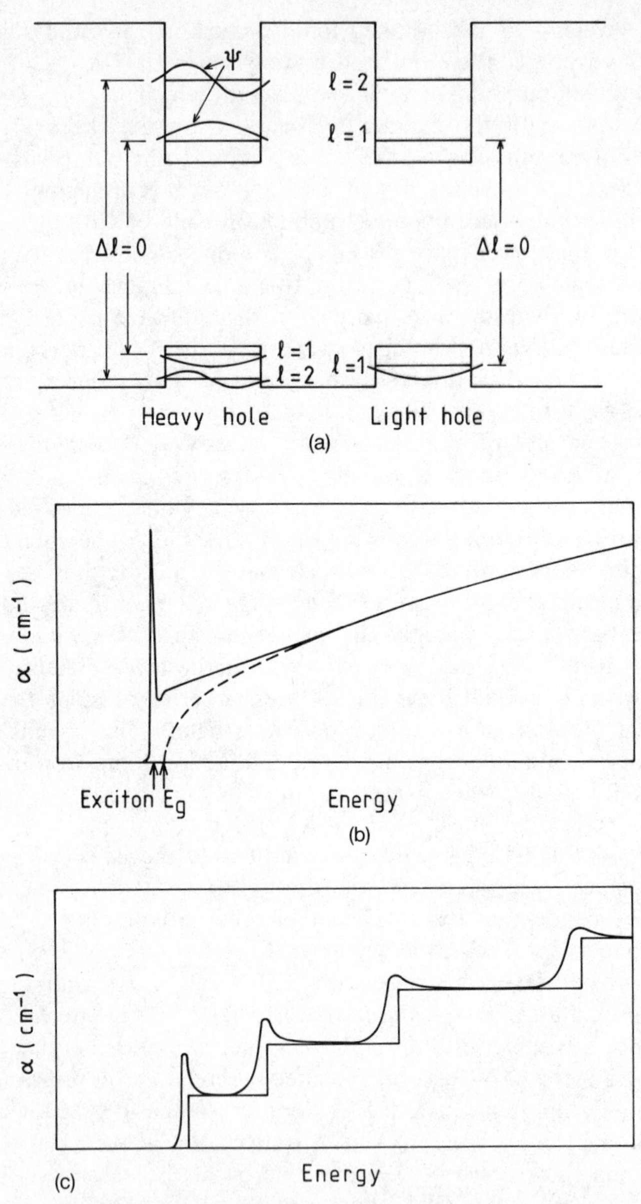

Fig. 7.2. (a) The band diagrams for a quantum well, indicating allowed optical transitions between the confined levels in the conduction band and the confined levels associated with the heavy (left-hand side) and light (right-hand side) hole valence bands; l labels the confined levels in each well as in Fig. 7.1c. Only transitions between levels with the same l are possible. (b) The absorption coefficient α as a function of photon energy for a bulk GaAs crystals, showing the

sets of confined levels derived from the valence band edge, one associated with m^*_{hh} and the other with m^*_{lh}. The other parameters determining the position of confined levels, the barrier height and the well width are, of course, the same in both cases. Accordingly, in Fig. 7.2a we have a sketch of two sets of levels near the valence band edge. The conduction band states appear twice simply for convenience; there is only one set of these states, since there is only one bulk conduction band near the band edge of GaAs.

7.2. Intrinsic optical properties of GaAs quantum wells

Let us now consider what happens when a quantum well is exposed to electromagnetic radiation. Light of frequency ω applied to a bulk crystal of GaAs can be absorbed by valence electrons when the photon energy $\hbar\omega$ is larger or equal to the magnitude of the forbidden gap. We have shown that the effect of confinement changes the energy of conduction and valence states in such a way that the uppermost valence state available in our quantum well system lies below the top of the valence band of bulk GaAs. The lowest conduction level lies above the bottom of the GaAs conduction band (as indicated in Fig. 7.1b). Consequently, in the quantum well structure, the energy needed in order to take an electron from the valence band into the conduction band is larger than the bulk gap of GaAs, i.e.

$$E_g(\text{Quantum well}) = E_g(\text{GaAs}) + E_{e1} + E_{hh1}, \tag{7.1}$$

where E_{e1} and E_{hh1} are the energies of the lowest confined electron and heavy hole states, respectively. These energies are positive because they are measured from the bottom of the corresponding confining well. The allowed transitions across the forbidden gap are indicated in Fig. 7.2a. The rule is that jumps are only possible between states with the same quantum number l, which we used to label the confined states in Fig. 7.1. One set of transitions is expected for electrons from the heavy hole valence band and one from the light hole band. Thus, in the structure

---

exciton line just below the onset of the band-to-band transitions across the forbidden gap E_g. The broken line shows the form of the bulk density of states ρ as a function of energy to demonstrate that α due to band-to-band transitions is propotional to ρ. (c) The absorption coefficient α for a quantum well structure; α can be seen to follow the step-like character of the density of states for two-dimensional systems shown under the absorption curve. The sharp feature near the edge of each step is the exciton contribution to α.

shown in Fig. 7.2a an optical experiment should yield three sharp lines whose energies can be predicted from an expression such as eqn (7.1), which determines the energy of the jump between the lowest states (the smallest photon energy at which a transition is possible); for each transition we have to use the relevant energies of the confined levels.

In order to explain the rule $\Delta l = 0$, we would have to invoke the symmetry properties of the bulk bands in some detail. However, only the principal transition across the gap, which is between the lowest lying states with $l = 1$, is of practical interest. It is easy to understand that this transition determines the threshold for absorption and emission in our quantum well, and we shall therefore confine our attention to it. The most important prediction of our theory is that this threshold energy, which is given by eqn (7.1), depends on the confinement effect. If we choose a particular aluminium concentration in the alloy material, we alter the conduction and valence band offset. As a result, the position of the confined levels is changed and with it the band gap of the quantum well structure. We can also alter the band gap by altering the well width. It follows that we can use the quantum well parameters to tune the optical properties of the system.

7.3. Extrinsic optical properties of GaAs quantum wells

In Chapter 4, we considered briefly the formation of a very special 'particle' peculiar to semiconductor materials, called the exciton. We described it as an electron–hole pair bound together by the attractive Coulomb force familiar from elementary electrostatics. The energy levels introduced into the forbidden gap as a result of the formation of the exciton resemble a hydrogenic series analogous to that for chemical impurities (see end of Section 4.4).

Excitons are normally created by exciting electrons from the valence band into the conduction band. In an experiment, this can be achieved by shining a laser beam of energy higher than the band gap on to the material. The electrons lifted into the conduction band by the beam rapidly relax to the bottom of the band, disposing of their kinetic energy in collisions with lattice vibrational waves (we say that they are emitting phonons). This process is very fast (of the order of picoseconds). Once at the bottom of the band, electrons find holes and form excitonic states that collapse after a lifetime of about a few nanoseconds: in the process of exciton annihilation, the electron emits a photon and jumps into the empty slot in the valence band (annihilating the hole). From our account of the electronic structure of a quantum well near the band edges, the same can happen there. However, in the quantum well system, the exciton energy (the depth of the excitonic level in the gap) must be measured from the band edge of the quantum well system. Hence the energy of the photon emitted upon the exciton dissolution is

$$\hbar\omega = E_g(\text{Quantum well}) - |E_1^{ex}|, \tag{7.2}$$

where E_1^{ex} is the exciton energy evaluated in Chapter 4 for bulk crystals (i.e. with the bulk crystal effective masses and dielectric constant ε) and E_g(Quantum well) depends on the band gap of bulk GaAs and the energy of the lowest confined conduction and valence levels eqn (7.1).

We know that in a bulk crystal of GaAs the radius of the exciton orbit is very large (>100 Å) because of the small electron effective mass. This means that even if we position the centre of the exciton orbit in the middle of a moderately thick well ($a = 100$ Å), the excitonic orbit extends well outside the GaAs layer. The simplest way to account for the interaction of the exciton with the step-like potential at the interface is to assume that the barrier there is infinitely high. Then the amplitude of the exciton wave function must be zero at the interface and inside the barrier material. The localization of the wave function is therefore considerably increased and both the electron and the hole spend more time closer to each other. The effect of the Coulomb potential binding the pair together is greatly increased because both particles spend more time near the centre of the orbit where the potential is more attractive. It follows that the net effect of the presence of the band offset is to increase the exciton binding energy. Detailed calculations show that the exciton energy can be increased due to the effect of confinement by as much as 300%.

Although this increase in exciton binding energy in GaAs wells amounts to only about 6 meV and constitutes a very small fraction of the band gap energy, it has important practical implications. Let us consider the optical spectrum of a bulk GaAs. The excitonic transition, in which a photon is emitted or absorbed at energy below the band gap energy, appears as a sharp peak in the optical spectrum. The band-to-band optical transitions that follow from their threshold at the bulk band gap energy appear in the form of a broad band of slowly increasing strength that roughly follows the increase in the bulk density of states away from the band edges (Fig. 7.2b). The fact that in bulk GaAs the exciton energy is only a few millielectron-volts means that even at temperatures well below 300 K the chances of observing the excitonic emission are very small. This is because the probability that the exciton is thermally ionized is given by the Boltzmann factor $\exp(-|E_1^{ex}|/k_B t)$, where $k_B T$ is about 26 meV at 300 K, so that the excitons are certain to be ionized. The increase in the exciton binding energy due to the confining quantum well potential greatly enhances the number of excitons, at least at lower temperature, and increases the optical signal. At low (liquid-helium) temperatures, the exciton-related emission (luminescence) in high-quality quantum well samples gives rise to a characteristic set of peaks (Fig. 7.2c). Although the most likely transition is due to the dissociation of the exciton formed by an electron from the lowest conduction sub-band and a hole from the uppermost valence sub-band, excitons can also be formed under favourable conditions by electrons and holes from other sub-bands following the selection rule $\Delta l = 0$ given in the preceding section. The separation between the observed peaks can be used to determine the position of the individual sub-bands empirically.

The sharpness and intensity of the exciton luminescence is a measure of the sample quality. The problem is the same as in bulk crystals. If the material is contaminated by defects, electrons and holes can be trapped in the potential wells associated with the presence of these defects, and only a small percentage of them are capable of forming excitons and emitting photons at the energy predicted by eqn (7.2). The quality of the GaAs layer in quantum well structures is always superior to the quality of the alloy forming the barrier. This is because it is more difficult to control the purity of the alloy material. Much of the unwanted trapping takes place at the interfaces, and the strength of the luminescence signal is therefore a measure of the interface contamination.

In GaAs quantum well structures, the line width of the exciton luminescence may also be a measure of the lack of uniformity of the interfaces. For example, if the interface contains steps, the well width 'seen' by an exciton whose orbit radius exceeds the well boundary appears to be smaller for some excitons and larger for others. We have argued that the exciton energy that enters eqn (7.2) actually depends on the well width. Consequently, in such structures the observed luminescence line is broadened or contains several closely spaced peaks, depending on the nature of the steps.

The effect of the confining barrier on the electronic structure of shallow chemical impurities in quantum wells is very similar to that we described for excitons. The presence of a single donor or acceptor impurity in an intrinsic bulk semiconductor leads to new levels, which form a hydrogenic series and which lie close to the conduction or valence band edge of the crystal. Such impurities are commonly used to dope semiconductors. If the radius of the orbit of the impurity state is large, the energy of the impurity levels will be increased in quantum wells. The magnitude of this enhancement depends on the distance of the impurity from the interface and on the well width. In general, the smaller the distance between the impurity site and the interfaces, the larger the localization of the wave function and consequently the larger the increase in the binding energy. This relationship is illustrated in Fig. 7.3. When the impurity sits very close to the interface, the simple notion of impenetrability of the barrier

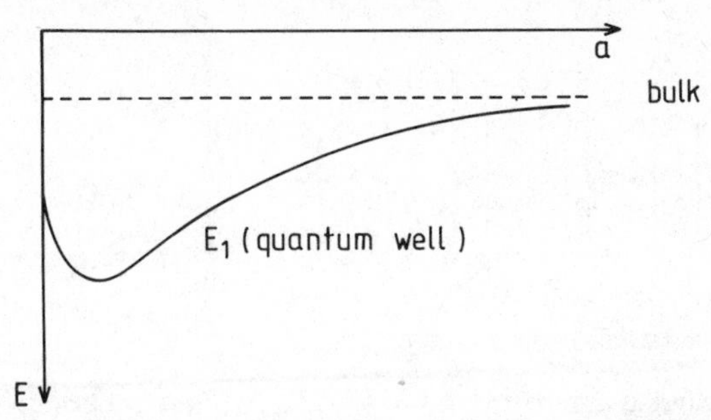

Fig. 7.3. The variation of the ground state energy E_1 measured with respect to the conduction band edge of the quantum well system (and treated as a negative quantity) of an impurity in a quantum well of width a. In the limit of large a, E_1 must converge to the bulk value (ΔE_1 given in Chapter 4), which is shown by the broken line.

breaks down and cannot be extended right up to the interface. Detailed calculations confirm this intuitive view and show that the binding energy reaches a peak at a certain critical distance from the interface and then drops slightly as the impurity site moves further towards the interface.

7.4. GaAs multi-quantum well structures and superlattices

Let us now consider a system consisting of a large number of alternating layers of GaAs and $Ga_{1-x}Al_xAs$. When the alloy layers are sufficiently thick (>100 Å), the tails of the wave functions associated with the states confined in the individual GaAs wells do not penetrate far enough into the barrier for them to overlap. Since the square of the wave function represents the probability of finding an electron or hole, the absence of significant overlap between adjacent wave functions means that carriers in such states are not free to move from well to well, and so remain localized. The structures in which the separation between wells is so large that each well may be treated as an independent (isolated) entity are called multi-quantum well (MQW) structures. The electronic properties of these structures are obtained simply as a superposition of the result we obtained for a single well, and there are no new levels.

When the alloy layers are made more narrow, the wave functions of the lowest states in the adjacent wells begin to overlap, as indicated on the left-hand side of Fig. 7.4. The solid line in each well indicates the position of the lowest confined state, E_1, obtained in the MQW limit. Because of the finite overlap, an electron in such a system is free to travel from one well to another. This is a situation familiar from Chapter 3,

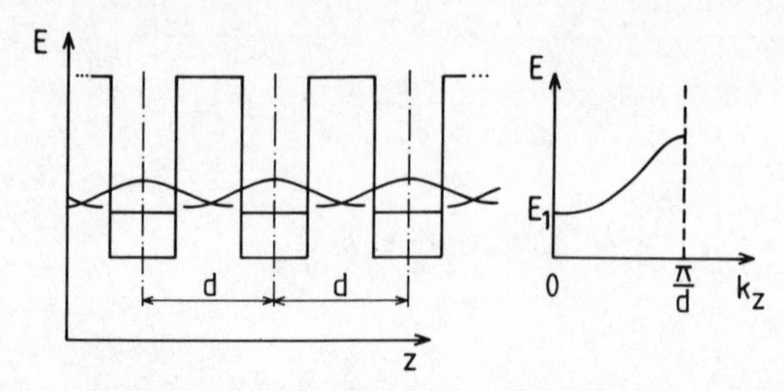

Fig. 7.4. The band diagram for a GaAs–$Ga_{0.7}Al_{0.3}As$ superlattice of period d. The overlap of the wave functions associated with adjacent wells, in the direction z perpendicular to the interface, is also indicated in the left-hand diagram. On the right we see the broadening of the lowest localized state E_1 of the single well into a band; π/d is the Brillouin zone boundary of this superlattice.

where we also considered a one-dimensional crystal consisting of weakly overlapping states localized at each lattice site. The only difference is that there we considered a chain of atoms located at regular intervals from each other. We first considered the separation between these atoms to be so large that the atoms could be regarded as independent. We associated each atom with a localized (atomic) wave function that would be obtained by solving the Schrödinger equation peculiar to the free atom in question. Then we brought the atoms closer together so that these atomic wave functions could overlap. We found that, irrespective of the details of the wave function and potential representing the atoms, we could predict that the original discrete (free-atom) energy levels become broadened into bands of allowed energies.

In this section, we can follow the same procedure to describe the solution of the Schrödinger equation for the periodic potential in Fig. 7.4, assuming that the barriers are chosen so that the overlap is finite but small enough for the tight-binding approximation to remain valid. In place of the atoms we have, as a point of departure, the square well potentials representing the GaAs quantum wells, and the wave functions described above, which are localized at each well. Also, we must replace the free-electron mass used in Chapter 3 by the effective mass from the relevant (conduction or valence) band edge of GaAs. As before, we shall ignore the small difference in effective masses of GaAs and $Ga_{1-x}Al_xAs$. The separation d between midpoints of adjacent GaAs wells (Fig. 7.4) replaces the lattice constant of our one-dimensional lattice of Chapter 3. Such a periodic structure of quantum wells is called a *superlattice* and d is the superlattice period. The energy as a function of wave vector in the direction of the superlattice axis (z) is given by eqn (3.15), where the atom parameters are replaced by those of the isolated quantum well, and the resulting dispersion can be drawn in the one-dimensional Brillouin zone whose boundary lies at $k = \pi/d$ and $-\pi/d$ (see the right-hand side of Fig. 7.4).

The model of the superlattice electronic structure pictured in Fig. 7.4 is based on the assumption that the separation between adjacent wells is small enough for the wave function overlap to be finite. However, by invoking the tight-binding theory of Chapter 3, we confined our interest to the structure in which the overlap may be regarded as a small correction. Then the band width is small and, apart from this broadening, the electronic structure still resembles that of a MQW system. If the width of the alloy layers is reduced further, the band width becomes comparable to or even larger than the separation of the levels of an isolated quantum well and the tight-binding model is no longer applicable. Instead, we must use the nearly-free-electron scheme of Chapter 2. Again, the theory developed there and in Appendix 3 is directly applicable to our superlattice problem.

Following the results of Chapter 2, the effect of a periodic potential consisting of steps whose height is equal to the magnitude of the conduction band offset is to divide the parabolic band $(\hbar k)^2/2m_e^*$, representing the bottom of the conduction band of GaAs (see the continuous parabola in Fig. 7.5) into a series of minibands separated by

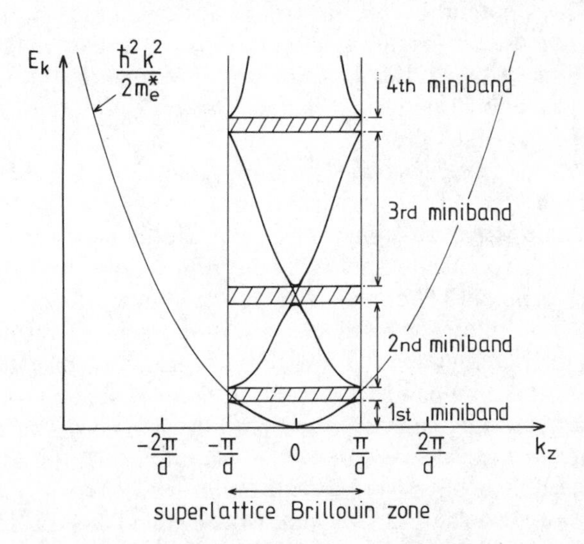

Fig. 7.5. The conduction miniband structure of a quantum well superlattice of period d, in the reduced-zone representation developed in the framework of the nearly-free-electron model in Chapter 2.

minigaps at the Brillouin zone boundaries (i.e. at $k = \pi/d$, $2\pi/d$, etc.). We can fold the minibands into the first Brillouin zone of length $2\pi/d$, and picture the electronic structure of our superlattice at the bottom of the conduction band in the reduced zone representation.

From a formal (mathematical) point of view, there is no difference between what we are doing here with the periodic quantum well potential and what we did in Chapter 2 with a general periodic crystal potential. However, the input parameters have different meanings and magnitudes. For instance, the Brillouin zone of the superlattice is much smaller since d is much greater than the lattice constant of bulk GaAs (hence the label 'minizone' or 'small Brillouin zone' is often used to distinguish the superlattice Brillouin zone from the Brillouin zone of the bulk crystal). The magnitude of the forbidden gaps can be obtained by evaluating the Fourier components V_g of the periodic potential at the superlattice reciprocal vectors of length $g = 2\pi/d$, $4\pi/d$, etc. In the simple nearly-free-electron (one-dimensional) model of Chapter 2, the forbidden gap is $2|V_g|$, where V_g is the Fourier component of the periodic potential evaluated at g. While in Chapter 2 we derived this result without any reference to an explicit form of the crystal potential—assuming that such a potential is too difficult to set up in a realistic fashion—here we actually have at our disposal a very good quantiative form of the periodic potential, since the barrier height is either the conduction or valence band offset. This potential is so simple that the magnitude of the gaps can be obtained numerically in a straightforward manner. In fact, in Appendix 3, we sketch the derivation of a general solution of the Schrödinger equation for the special case involving a step-like periodic potential. This so-called Kronig–Penney solution provides an analytical formula for energy-versus-wave vector dispersion relations pictured schematically in Fig. 7.5.

The electronic structure shown in Fig. 7.5 reflects the changes in the energy of an electron at the bottom of the conduction band of GaAs due to the periodic potential representing the difference between the alternating GaAs and alloy semiconductors. As the general results given in Chapter 2 and in Appendix 3 show, the magnitude of the minigaps and the width of the minibands depend on the conduction band offset, effective mass, and the well and barrier widths. (If we desire to evaluate the minibands at the valence band edge, we must perform an analogous calculation, using the effective mass for either the heavy or the light hole valence band of GaAs and the relevant valence band offset for the barrier height). Such calculations give us the energy versus k in the direction perpendicular to the interface. To obtain the total energy, we shall assume that the motion in the direction parallel to the interface plane remains the same as in bulk GaAs. This is the same assumption we have made earlier in this chapter and the expression for the total energy can be obtained following the procedure described for isolated wells in Fig. 7.1c.

7.5. $Ga_{1-x}Al_xAs$ sawtooth superlattice

In Section 7.4 we established a simple computational scheme for modelling the electronic structure of square well superlattices. As in the case of isolated quantum wells, the usefulness of such structures lies in the opportunity to 'engineer' the electronic structure. For example, we can change the forbidden gap and make the material suitable for certain applications in optoelectronics in a particular range of the optical spectrum. In comparison with MQW structure, the superlattice offers an additional degree of flexibility in that it is now also possible to choose the superlattice parameters so as to alter the band curvatures.

It is borne in mind that it might be of interest to look at other superlattice structures. The simplest way to deviate from the square well arrangement discussed in Sections 7.1–7.4 is to create a 'sawtooth' potential profile shown in Fig. 7.6. This can be achieved by growing the sample in such a way that, starting from pure GaAs, each subsequent atomic layer deposited in the growth chamber contains a slightly higher percentage of aluminium. When the desired aluminium concentration is reached, the growth cycle is restarted beginning with pure GaAs. We end up with layers consisting of material with graded band gap (and band offset). Figure 7.6 shows the potential profile seen by electrons at the

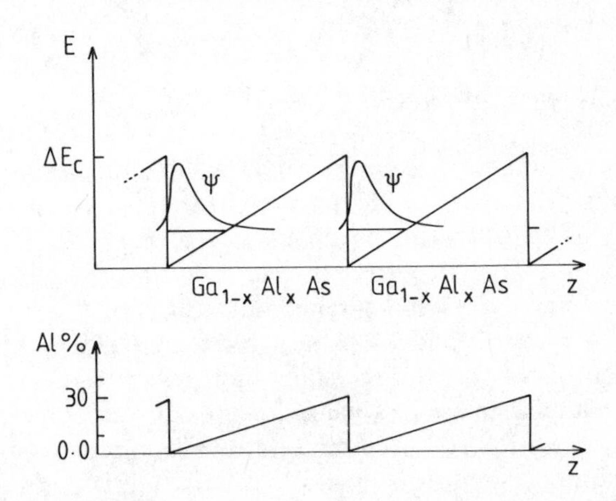

Fig. 7.6. The conduction band diagram for a sawtooth superlattice, obtained by grading the GaAs layers with varying fraction of aluminium. The latter is indicated in the lower diagram.

bottom of the conduction band of one such sawtooth superlattice. The form of the electronic structure of the sawtooth superlattice can also be understood in terms of the model outlined in Section 7.4. If the width of each graded layer is large, the wave function of the confined levels will be localized mainly in the regions of deeper potential, as illustrated in the figure, and the contact between adjacent triangular wells will be weak. In a structure with narrow layers, the overlap of the adjacent wave functions is strong and we must use the nearly-free-electron model to calculate the width of the minibands. As in the case of square well superlattices, we expect the width of the higher minibands lying closer to the top of the confining barrier to be larger, until near and above the top of the barrier we encounter a continuum of bulk-like states. This is well in keeping with our understanding of the nature of the wave functions that we obtained for isolated wells and whose overlap determines the band width. We noted that a state lying near the bottom of the well is more confined (localized) in the well than the states lying above it.

In Fig. 7.6, only the potential barrier peculiar to the conduction band is shown. An analogous diagram can be drawn for the calculation of confined levels near the valence band edge, where the height of the confining barrier of the graded structure depends on the valence band offset. Note that the grading separates electrons from holes in that the deep ends of the triangular wells for the calculation of conduction and valence levels, i.e. the regions where the corresponding confined states are localized, lie at the opposite ends of the graded layer. This property of the sawtooth structure will be invoked in connection with device applications.

7.6. Doping superlattices

A superlattice structure analogous to that outlined in Section 7.4 can be created by doping a semiconductor crystal (e.g. GaAs, Si) with donors and acceptors as indicated in the upper part of Fig. 7.7. The impurities are introduced in a highly controlled fashion so that the material consists of alternating layers of n- and p-type character. For this reason doping superlattices are often called nipi structures. Since the impurities are randomly distributed in the crystal, the layers do not have such well-defined boundaries as the quantum well structures discussed above. Also, the concentration of the impurity atoms is limited by the solubility of the species in question. This means that very thin layers are difficult to achieve by this technique.

In equilibrium, electrons are transferred from the donor layers to fill the empty (hole) states in the acceptor layers. They leave behind

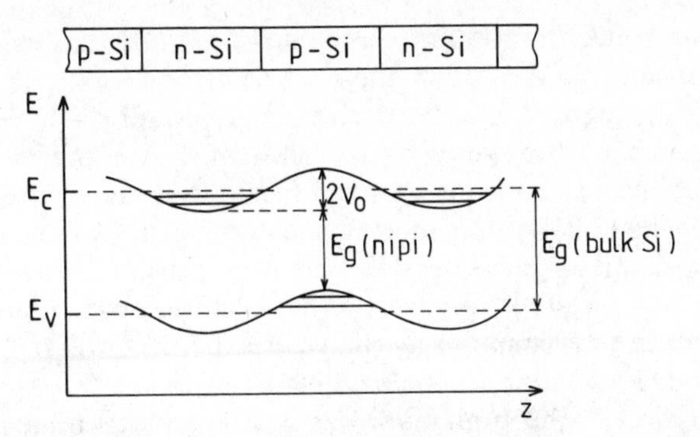

Fig. 7.7. The periodic space charge potential of amplitude $2V_0$ obtained by doping a silicon crystal by donors and acceptors, as illustrated in the upper diagram. The fundamental band gap E_g of this nipi structure and that of bulk silicon are indicated. E_c and E_v are the conduction and valence band edges of bulk silicon. The confined levels in the parabolic wells are indicated by solid horizontal lines.

positively charged donors. The acceptor layers are negatively charged. This gives rise to an electrostatic potential familiar from our account of the electronic structure of p–n junctions. The periodic nature of this potential makes the problem amenable to the same treatment as that implemented in the case of the square well structures. A rough sketch of the electrostatic potential superimposed upon the bulk band structure of the host crystal, represented here by the conduction and valence band edges at E_c and E_v, respectively, is shown in the lower part of Fig. 7.7. Normally, we expect the individual doped layers to be about 100–300 Å thick, so that the deepest confined levels associated with the adjacent potential minima are quite independent of each other. The position of the confined energy levels is also indicated in Fig. 7.7 by solid horizontal lines in each 'well'. As this figure suggests, the minima of the potential determining the wells for states derived from the conduction band are spatially separated from the maxima that determine the wells for confined levels originating in the valence band. This spatial separation of electron and hole states is a characteristic property of doping superlattices that distinguishes them from the GaAs quantum well structures.

The precise position of the confined levels lying below the bulk conduction band edge can be obtained from the Schrödinger equation containing the space charge potential:

$$\left\{-\frac{\hbar^2}{2m_e^*}\frac{\mathrm{d}^2}{\mathrm{d}z^2}+\tfrac{1}{2}qz^2\right\}\psi = E\psi. \tag{7.3}$$

Here q is a parameter that depends on the width of the doped layer and on the doping concentration (see Problem 7.5); m_e^* is the effective mass of the conduction band edge of the host semiconductor. The parabolic potential used in eqn (7.3) is particularly simple and remains valid only near the bottom of the well ($z = 0$). However, it is sufficient for the calculation of the lowest-lying confined levels that are of main interest. The equation in (7.3) is one of the 'standard' problems in elementary quantum mechanics, since it describes the properties of a simple one-dimensional harmonic oscillator. The solutions, which are described in great detail in any quantum-mechanics textbook, are $E_n = \hbar\omega(n + \frac{1}{2})$, where $n = 0, 1, 2, \ldots$ and $\omega = (q/m_e^*)^{1/2}$.

The calculation of the confined levels near the valence band edge can also be obtained from eqn (7.3) if m_e^* is replaced by either m_{hh}^* or m_{lh}^*. Note that the separation of the energies for the harmonic oscillator problem follows a different pattern from that in square wells.

If we reduce significantly the width of the doped layers, we can make the wave function from the adjacent conduction (and valence) parabolic wells overlap. The discrete electronic levels obtained from eqn (7.3) are then broadened into bands. We can follow the procedure described in the above sections to compute the dispersion relations. In this case the potential is again simple enough to make such a calculation possible with the means at our disposal. However, as pointed out at the beginning of this section, the usefulness of this type of structure lies in applications of thicker layers, and the wells may be viewed as independent of each other. Additional tunability is achieved by changing the doping concentration that controls the strength of the space charge potential (V_0 in Fig. 7.7).

7.7. Limits of tunability of semiconductor superlattices

The quantum wells and superlattices considered above are the archetypal semiconductor microstructures. They were also the first well-characterized microstructures suitable for practical applications, and served as a vehicle for establishing the key concepts of band structure engineering. Once the potential of these microstructures had been identified, and the objectives of the materials scientists had been formulated in terms of the concept of tunability, other structures were sought in order to satisfy a variety of needs in microelectronics. From a practical point of view, we always seek materials whose lattice constant is very similar. The form and strength of the confinement effect, and its influence upon the electronic structure we want to 'tune', depends largely on the magnitude of the band offset and on the bulk gaps of the constituents.

The relationship between these parameters is best appreciated if we

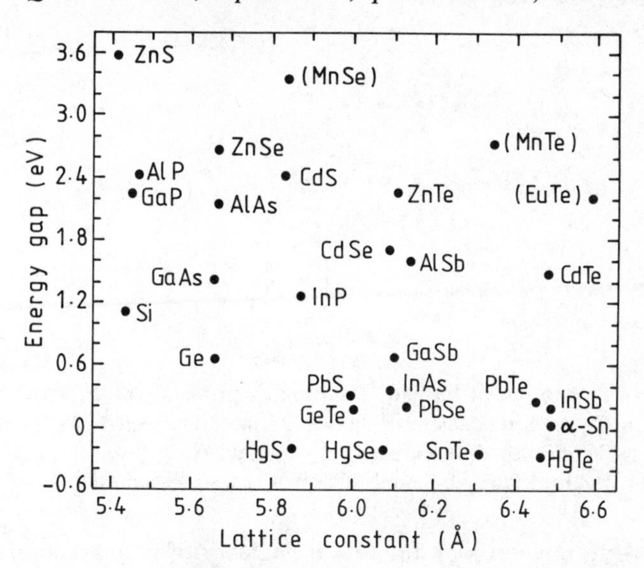

Fig. 7.8. The fundamental gap at room temperature of some important semiconductor crystals as a function of the lattice constant.

plot the fundamental (bulk) gaps of some key materials as a function of the lattice constant. This plot is shown in Fig. 7.8. Clearly, materials lying on a vertical line are lattice matched. Other combinations can be thought of if we are prepared to use ternary and quaternary alloys. For example, we considered in Chapter 5 alloys of InP, GaAs, GaP, and AlAs in which the material fractions were chosen so as to produce a desirable band gap and a lattice constant that matched the required substrate (e.g. GaAs or InP). Similar opportunities exist in the choice of partners for quantum well structures, although there are now more parameters to choose from because of the effect of confinement. Since alloy semiconductors are generally of poorer quality, the tendency is to take advantage of the confinement effect first. Quantum well structures as varied as InP–$Ga_{1-x}In_xAs$, GaSb–InAs, and CdTe–HgTe have been exploited.

In all of the structures we have so far considered, the difference between the lattice constants of the constituent semiconductors has been of order 0.01 Å or less, i.e. a small fraction of 1 per cent. However, it has been shown that it is possible to grow high-quality layers of materials when the lattice mismatch is as large as several per cent. This situation is illustrated in Fig. 7.9, which shows schematically two bulk lattices of different separation, i.e. before they are joined to form a heterojunction structure. The most notable example of practical importance is the Si–Ge superlattice. The difference in the lattice constant means that, for

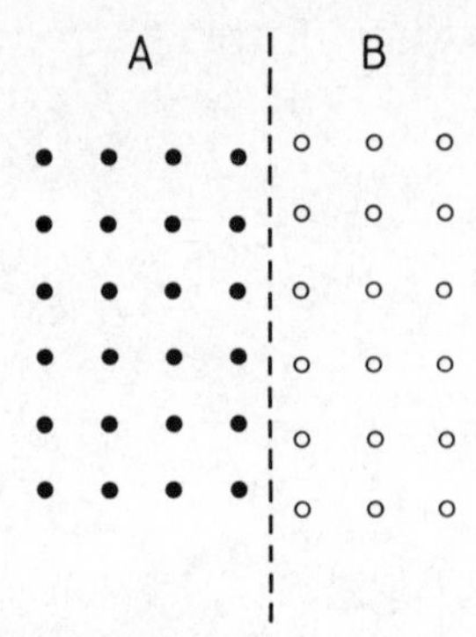

Fig. 7.9. An illustration of the lattice mismatch problem, which arises when an attempt is made to form a heterojunction of materials A and B whose lattice separation is significantly different. If B is grown on A, its lattice must contract in the interface plane and this gives rise to a strain field.

instance, the atoms forming the first monolayer of germanium deposited on the surface of a thick silicon wafer are forced to assume, in the direction parallel to the interface, the spacings between atoms characteristic of bulk silicon. If we imagine that in a macroscopic crystal of germanium atoms are held together by strong springs connecting the nearest neighbours, then it is obvious that the deposition on silicon compresses the germanium lattice in the plane of the interface. The energy stored in the springs as a result of this compression can only be released by forcing the atomic separation in the deposited germanium layers to expand in the direction perpendicular to the interface. When many layers of germanium have been deposited, the separation of germanium atoms is likely to converge to the equilibrium lattice separation of a macroscopic germanium crystal. However, if our deposited layer of germanium is thin enough, and we grow on top of it silicon again, the separation between the germanium atoms remains uniform. We end up with a strained layer of germanium whose lattice constant in the direction perpendicular to the interface is increased and in the parallel direction is reduced compared to the equilibrium bulk germanium value. The strain in this layer may, in fact, improve the quality of the material, since it provides a force that can push out dislocations.

We can use the band offset estimated in Chapter 6 to set up a band diagram of a Si–Ge superlattice. However, we must recognize that the strain due to the lattice mismatch also contributes to the potential seen by electrons and holes in this structure and alters the band line-up. Since our theory, on which the band offset of Chapter 6 is based, does not take the strain effect into account, the magnitude of the valence band offset ΔE_v given in Chapter 5 is only a crude approximation. Notice that the band

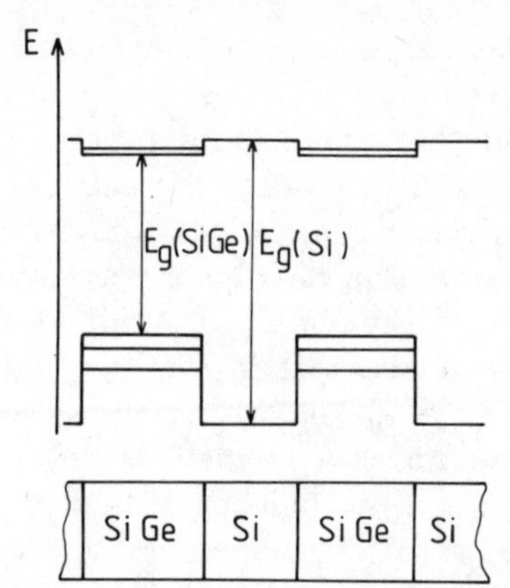

Fig. 7.10. The band diagram for a quantum well structure made of alternating Si and SiGe layers. The horizontal lines show the approximate positions of confined levels in the wells. (SiGe is an alloy containing 50% Si and 50% Ge.)

gap of germanium is smaller than the band gap of silicon and the difference is such that a small error in ΔE_c may alter the very sign of the step that determines whether electrons are confined in germanium or in silicon layers. The magnitude of the band offset can be manipulated further by employing an alloy of germanium and silicon instead of pure germanium. The band diagram for such a structure, derived from the offset given in Chapter 6, is sketched in Fig. 7.10. However, experiments indicate that the valence band offset is in fact larger, enough to shift the bottom of the conduction band well into the silicon layer.

The band gap obtained in this structure can be achieved in quantum well systems consisting of other, much better lattice-matched, partners. The advantage of silicon-based structure is that it is compatible with other silicon-based devices. Furthermore, both silicon and germanium are well-characterized and technologically well-understood materials.

A glance at the Periodic Table suggests that there are almost limitless opportunities for designing new structures. However, the potential advantages of a novel artificial structure predicted theoretically (e.g. from electronic structure calculations) must also be considered in terms of the cost and time required to develop and manufacture such a new material.

7.8. Quantum wires and dots

Semiconductor quantum well structures described in the above sections exploit the confinement that occurs in the direction perpendicular to the interfaces. This one-dimensional confinement effect is used to engineer the electronic properties of the material. The number of the adjustable parameters available in this engineering process can be increased by creating analogous structures in which the carriers are confined in two and three dimensions. Such structures are often called quantum wires and dots, respectively. We can visualize a quantum wire as a long rectangular rod of GaAs embedded in, say, $Ga_{1-x}Al_xAs$. If we lay this rod along the z direction, we obtain confinement in the x and y directions, with the relevant barriers constructed in the same way as for quantum wells (Fig. 7.11). For example, we have a barrier of height ΔE_c that prevents electrons at the bottom of the conduction band of GaAs from escaping into the alloy in the x and y directions. We can assume that the confinement effects in the x and y directions are independent processes. In that case, the model leading to detailed description of the confined levels is very simple. To obtain the total electron energy, it is sufficient to make a superposition of the energies obtained for confinement in one dimension. For instance, if (for a given value of well width a and barrier height ΔE_c) the ground state energy measured from the bottom of the GaAs conduction band edge in the one-dimensional case is E_1, then in the case of two-dimensional confinement, peculiar to the quantum wire in Fig. 7.11, that energy is $2E_1$. This means that by increasing the dimensionality of the confinement effect (or, we might also say, by decreasing the dimensionality of the GaAs layer from 2 to 1), we lift the lowest confined level further from the bottom of the well. Since the barrier height is independent of dimensionality, the position of a confined level in a quantum wire is twice as sensitive to the well width as that in a quantum well.

It is quite obvious that we can go one step further and consider a

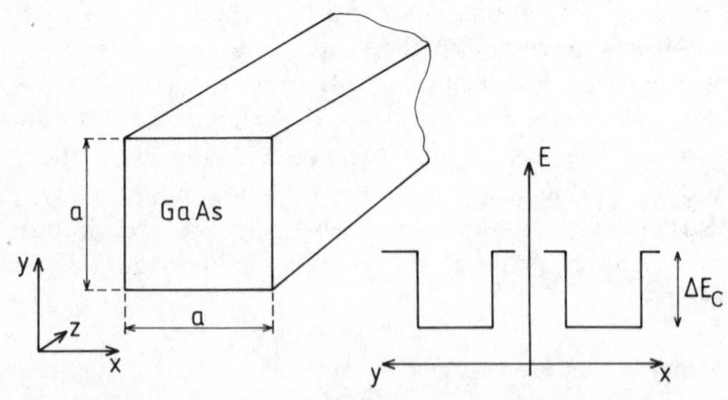

Fig. 7.11. A quantum wire of GaAs of cross section a^2. On the right is a sketch of the conduction band diagram for the calculation of confined levels. ΔE_c is the conduction band offset between GaAs and the $Ga_{1-x}Al_xAs$ alloy in which the wire is embedded.

cube—a quantum box or dot as it is often called—of GaAs embedded in a $Ga_{1-x}Al_xAs$ alloy material. We then have three-dimensional confinement, and the levels lie three times higher above the bottom of the conduction band of bulk GaAs compared to the quantum well case. The calculation of the level positions in this approximation is a straightforward extension of our procedure adopted for quantum wells and superlattices at the beginning of this chapter.

Problems

7.1. Find the change (in eV) of the band gap of a GaAs–$Ga_{0.7}Al_{0.3}As$ quantum well structure as a function of well width a where $0 < a < 200$ Å.

7.2. Find the contact potential for a GaAs–$Ga_{0.7}Al_{0.3}As$ heterojunction if the barrier material is doped with 10^{17} cm^{-3} donors ($T = 300$ K).

7.3. Consider a periodic potential representing the conduction band diagram of a GaAs–$Ga_{0.7}Al_{0.3}As$ superlattice. Assuming that the well width a is the same as the width of the barrier layer, find a such that the width of the lowest miniband is 100 meV. Justify the approximations you make.

7.4. Describe in qualitative terms the form of the change in the superlattice band diagram due to doping. In particular, show that as the well and barrier width decrease in magnitude, the space charge potential contribution to the barrier height also decreases.

7.5. Consider a doping superlattice in GaAs. Assume that the width a and impurity concentration s in the n- and p-type layers are the same. Find the amplitude $2V_0$ of the periodic space charge potential shown in Fig. 7.7 in terms of a and s. If $a = 200$ Å, determine s in cm^{-3} such that $4V_0 = \frac{1}{2}E_g$(GaAs).

7.6. Use the results of Problem 7.5 to estimate the position of the lowest confined level at the conduction and valence band edges and determine the band gap.

7.7. Calculate the position of the lowest valence (heavy hole) level in a GaAs–$Ga_{0.7}Al_{0.3}As$ quantum well of width $a = 50$ Å, a quantum wire of cross-sectional area a^2, and a quantum box of volume a^3.

7.8. What is the wavelength of the light absorbed in the structure with the largest gap that can be obtained with GaAs–$Ga_{0.7}Al_{0.3}As$ heterostructures at room temperature?

7.9. Recommend a well lattice-matched quantum well system based on a pair of II–VI semiconductors that could be used as an efficient emitter of blue light.

7.10. Recommend a lattice-matched quantum well structure consisting of group IV and II–V semiconductors that could be used to make infrared detectors.

7.11. Recommend an alloy and a quantum well structure based on an combination of GaP and GaAs that could be used to make an efficient emitter at low temperatures (i.e. a light-emitting diode) of red light.

8
Electron transport in microstructures

8.1. Transmission properties of a rectangular barrier

The most common structural unit, on which much of the physics of very small semiconductor systems depends, is a heterojunction. The band diagram representing a heterojunction is a step-like potential barrier. We have explored the changes in the electronic structure introduced by the barriers due to band discontinuities (offsets), in particular the existence of new quantum states with localized (confined) wave functions. We have also made a number of predictions concerning the tunability of optical (absorption and emission) spectra of these systems. In this chapter, we shall inquire about the ways these changes in the electronic structure affect transport properties of semiconductor microstructures.

In classical physics, a barrier is an impenetrable obstacle to a moving particle, unless we give the particle an energy so large that it can jump over the top. The effect of thermionic emission outlined in Chapter 4 is an example of such a situation. However, electrons are microscopic particles whose properties are governed by the laws of quantum mechanics, and the probability of finding an electron inside the barrier material is finite except when the barrier is infinitely high. This is the case even though the electron may lie well below the top of the barrier. The degree to which the electron penetrates into the barrier is measured by the magnitude of the squared modulus of the solution (wave function) of the stationary Schrödinger equation inside the barrier. If the barrier is thin, there is a finite probability of finding the particle on the other side. Thus, a current of electrons need not necessarily be stopped by a barrier of finite height and width. We say that electrons can tunnel through it.

Naturally, we want to apply these ideas to our heterojunction structures. Since these structures are represented by step-like potentials, we shall begin with a brief outline of tunnelling through a simple rectangular repulsive barrier. It will provide us with a basis for more detailed considerations peculiar to heterojunction structures.

A repulsive barrier of height V is shown in Fig. 8.1. If we put a free electron of mass m at an energy E on the left-hand side of the barrier, we

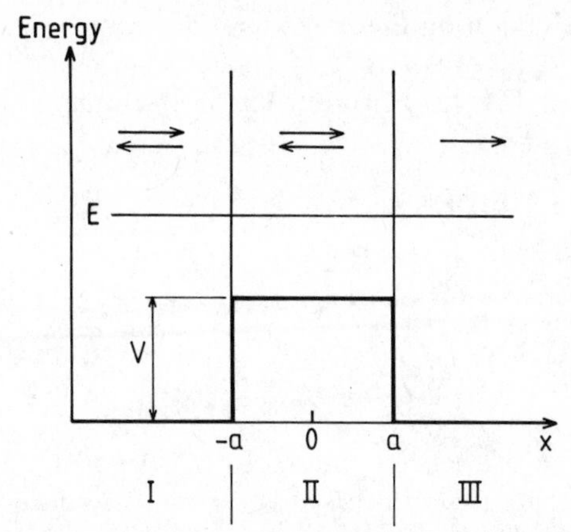

Fig. 8.1. A repulsive barrier of height V and width $2a$. The form of the wave function in domains I, II, and III is given in eqns (8.1)–(8.3). E shows the path of an electron moving at energy E.

want to know the probability T of finding it on the other side. The method for solving this problem is analogous to that adopted in Appendix 2, where we solve the one-dimensional Schrödinger equation for a particle in a box. Let us first consider the case in which the electron energy is larger than V, i.e. $E > V$. In region I of Fig. 8.1 we can write the wave function ϕ as

$$\phi_I = A \exp(ik_1x) + B \exp(-ik_1x); \qquad E = (\hbar k_1)^2/2m, \tag{8.1}$$

where the first term in the expression for ϕ_{I} represents the incoming wave and the second the reflected wave. Similarly, in regions II and III in Fig. 8.1 we have wave functions of the forms

$$\phi_{\mathrm{II}} = C \exp(ik_2x) + D \exp(-ik_2x); \qquad E - V = (\hbar k_2)^2/2m \tag{8.2}$$

and

$$\phi_{\mathrm{III}} = \mathrm{F} \exp(ik_1x); \qquad E = (\hbar k_1)^2/2m. \tag{8.3}$$

The reflection coefficient R, and the transmission coefficient T, are defined in terms of the ratios of the amplitudes of the incoming and reflected wave, and of the incoming and transmitted waves, respectively

$$R = |B/A|^2; \qquad T = |F/A|^2. \tag{8.4}$$

The consistency condition for T and R is

$$T + R = 1. \tag{8.5}$$

The amplitudes A, B, C, D and F required to determine R and T are obtained from the condition that the wave functions ϕ_{I}, ϕ_{II}, and ϕ_{III}, and their derivatives with respect to x, are continuous at the barrier boundaries $x = \pm a$. This leads to four linear algebraic equations for ratios B/A, C/A, D/A, and F/A:

$$\begin{aligned}
&\exp(-\mathrm{i}k_1 a) + (B/A)\exp(\mathrm{i}k_1 a) = (C/A)\exp(-\mathrm{i}k_2 a) + (D/A)\exp(\mathrm{i}k_2 a);\\
&k[\exp(-\mathrm{i}k_1 a) - (B/A)\exp(\mathrm{i}k_1 a)]\\
&\qquad = k_2[(C/A)\exp(-\mathrm{i}k_2 a) - (D/A)\exp(\mathrm{i}k_2 a)];\\
&(C/A)\exp(\mathrm{i}k_2 a) + (D/A)\exp(-\mathrm{i}k_2 a) = (F/A)\exp(\mathrm{i}k_1 a);\\
&k_2[(C/A)\exp(\mathrm{i}k_2 a) - (D/A)\exp(-\mathrm{i}k_2 a)] = (F/A)\exp(\mathrm{i}k_1 a). \quad (8.6)
\end{aligned}$$

Solving the last two of these equations for D/A and C/A as functions of F/A, and substituting into the first two, gives the solution for B/A and F/A. If we express k_1 and k_2 in terms of V and E, we obtain for the transmission coefficient,

$$\frac{1}{T} = \left|\frac{A}{F}\right|^2 = 1 + \frac{V^2 \sin^2(2k_2 a)}{4E(E-V)}; \qquad E > V \qquad (8.7)$$

For the case of $E < V$, we can still make use of the result in eqn (8.7) if we make a substution

$$\mathrm{i}k_2 = \beta; \qquad \frac{(\hbar\beta)^2}{2m} = V - E > 0. \qquad (8.8)$$

Since $\sin(\mathrm{i}y) = \mathrm{i}\sinh(y)$, eqn (8.7) becomes

$$\frac{1}{T} = 1 + \frac{V^2 \sinh^2(2\beta a)}{4E(V-E)}; \qquad E < V. \qquad (8.9)$$

The behaviour of the transmission coefficient T as a function of electron energy in units of the barrier height V is shown in Fig. 8.2. We can see that T increases rapidly with increasing energy and acquires significant magnitude well before it reaches the top of the barrier at $E = V$. This is well in keeping with our frequent references in foregoing chapters to the tail of the wave functions of confined levels that also penetrate into the barrier layers. The degree of such penetration in superlattice structures gives a quantitative meaning to the broadening of confined states into bands.

Our theory predicts not only that electrons with $E < V$ can penetrate the barrier (a notion familiar from our previous discussion of the electronic structure), but that the states of electrons with $E > V$ are also affected. In particular, from eqn (8.7) we can deduce that when the barrier width $2a$ becomes an integral number n of half-wavelengths $\lambda/2$, the barrier is transparent to the incident beam ($T = 1$ when $\sin(2k_2 a) = 0$

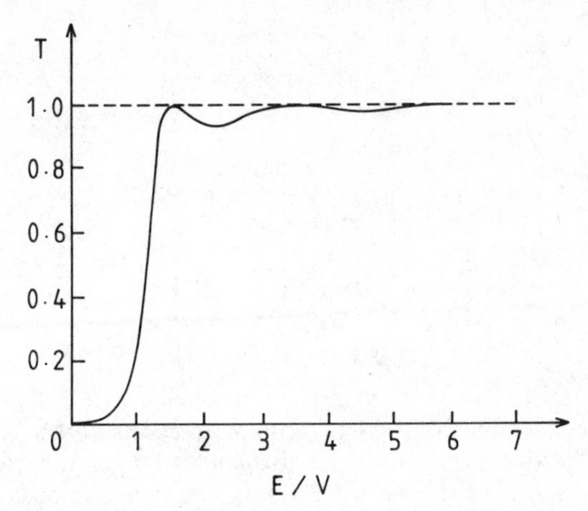

Fig. 8.2. The transmission coefficient T (defined in the text) for a repulsive barrier of height V. E is the energy of the travelling particle.

with $k_2 = 2\pi/\lambda$ such that $2ak_2 = n\pi$). If we write the condition for perfect transmission in terms of E and V, we find for $E > V$ that

$$E - V = \frac{(n\pi\hbar)^2}{8a^2m} = n^2E_1, \tag{8.10}$$

where E_1 is the ground-state energy of a one-dimensional square well of width $2a$, $n = 1, 2, \ldots$.

In our discussion of the electronic structure of quantum wells and superlattices, we also made use of step-like potentials considered here. We concentrated on the changes in the electronic structure of bulk materials (e.g. GaAs) introduced by such potentials at the conduction and valence band edges (on the levels lying deep in the well), i.e. in the range of energies $E < V$ in the notation used in this section. The result in eqn (8.10) suggests that in the direction perpendicular to the interface planes the electronic structure above the confining barrier is not entirely bulk-like and must be also affected by the band discontinuities.

In fact, there is no reason why the calculations based on the model of electronic structure developed Chapter 2 could not be carried out in the range of energies lying above the confining barriers, provided that the bulk band can still be regarded as parabolic at such higher energies (i.e. that the concept of effective mass remains approximately valid). The Kronig–Penney solution of Appendix 3 can also be directly applied to this energy range, and it can be verified that it predicts the existence of one or more 'confined' levels lying above the barrier. However, the wave function of such a state is localized mainly in the barrier region, as expected on intuitive grounds from eqn (8.10).

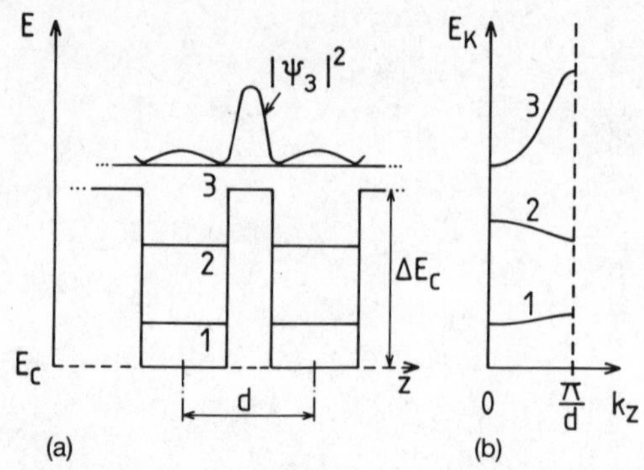

Fig. 8.3. (a) A conduction band diagram for a superlattice of period d, showing two wells whose barrier height is given by the conduction band offset ΔE_c. Levels 1 and 2 are the solutions of the Schrödinger equation for each isolated well, whose energies lie in the wells. (b) The dispersion that arises when there is an overlap between adjacent wave functions. This broadens the localized levels 1 and 2 into minibands 1 and 2. When solutions of the same equation are sought in the range of energies above the barrier, one again obtains, in the direction perpendicular to the interfaces, minigaps and minibands. The lowest such miniband (3) is indicated. Its wave function peaks in the barrier, as expected from the result in Fig. 8.2. Wave functions associated with bands 1 and 2 peak in the wells.

The form of this wave function and the expected dispersion relations are sketched in Fig. 8.3. The lowest two bands lie below the confining barrier (band offset) and are familiar from our earlier discussion of the electronic structure of GaAs–$Ga_{0.7}Al_{0.3}As$ superlattices. Band 3 is the new feature we were inspired to investigate after seeing the result in Fig. 8.2. This has interesting consequences for electron transport in microstructures in the direction perpendicular to the interfaces. In small systems, electrons are often 'hot', that is, they can gain energy, for example by being accelerated in the strong electric field prevailing in such systems. Such electrons travel in the system with energy much higher than the lowest conduction band minimum. Our theory predicts that if these electrons find their way into one of the higher bands above the barrier, they may be delayed in their flight and forced to spend more of their time in the barrier layers than they would have done had the electronic structure there been bulk-like. Because these bands lie above the barrier, in the range of energies where we expect a continuum of bulk quantum states, they are often referred to as resonances.

8.2. Tunnelling in heterojunctions

The results outlined in the preceding section can be applied directly to transmission (tunnelling) through intrinsic heterojunction barriers provided that m and V are replaced by the corresponding effective mass and band offset, respectively. However, in most situations of practical interest, the barrier potential is not a strictly rectangular step. Even though we have always assumed that the interface between two semiconductors is abrupt and free of irregularities, the effect of doping and the applied electric field destroy the step-like character of the barrier, which must be considered in order to describe electron motion. This does not alter the general conclusions concerning the physics of transmission and reflection processes, but it does mean that under such circumstances any quantitative predictions we might wish to make about these processes from equations for T and R in Section 8.1 would be quite unrealistic. We must, therefore, find a formalism that will enable us to employ more general forms of the potential.

In Chapter 6, we investigated the form of the potential barriers characteristic of the metal–semiconductor (Schottky barrier), MOS, and semiconductor–semiconductor (heterojunction) contacts, with and without an external electric field. The most common form of potential we encountered there resembled a triangular barrier.

The barrier shape in Fig. 8.4 could be obtained as a result of superposition of a step-like band discontinuity and an applied electric field $\mathscr{E}$ ($F = e\mathscr{E}$). On the left-hand side of Fig. 8.4, we have a certain density of electrons, with the uppermost occupied level at energy E_F. This could represent, for example, the conduction band of biased n-type GaAs forming heterojunction with $Ga_{0.7}Al_{0.3}As$. W indicates the separation in energy of the top of the barrier and the electron reservoir; x is perpendicular to the interface plane. We shall now generalize our treatment of tunnelling by exploiting one of the characteristic features of this problem, namely that under normal circumstances the barrier potential is slowly varying on the scale of the electron wavelength. Consider an electron propagating along x from the left at some energy E, where $0 < E < W$. The condition that potential $V(x)$ is slowly varying allows us to assume that inside the barrier the wave function $\phi(x)$

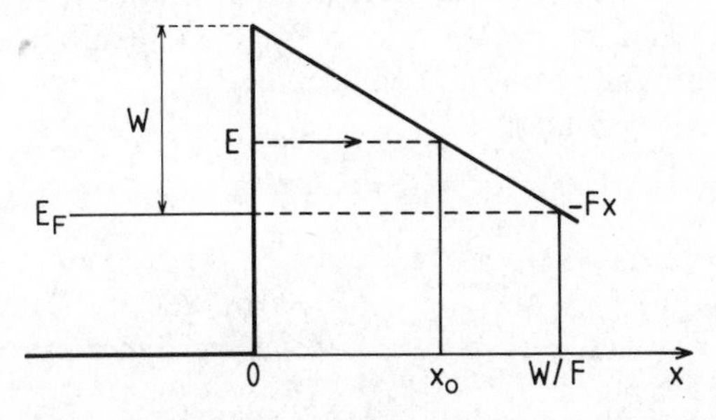

Fig. 8.4. A triangular barrier that might represent a heterojunction with an applied electric field F/e.

resembles a free-particle plane-wave solution

$$\phi(x) = A\exp(ikx) = A\exp(ipx/\hbar). \tag{8.11}$$

The effect of the barrier potential V is to alter the momentum p of the particle slightly as the particle propagates from $x = 0$ to the right, i.e. p is a slowly varying function $p(x)$ of x. The solution inside the barrier can be written in an approximate general form as

$$\phi(x) = A\exp(\mathrm{i}S(x)/\hbar). \tag{8.12}$$

The net change of momentum up to a point x_0 of our electron moving along the path from 0 towards the end point x_0 (e.g. at energy E indicated in Fig. 8.4) is $(p^2/2m = V - E)$,

$$J(x_0) = \int_0^{x_0} p(x)\,\mathrm{d}x = \int_0^{x_0} (2m[V(x) - E])^{1/2}\,\mathrm{d}x. \tag{8.13}$$

Hence, if the amplitude of ϕ is 1 at $x = 0$ then its amplitude at x must be $\exp(-J(x)/\hbar)$. The transmission coefficient is defined as the square of the ratio of the wave function amplitude at the entry and exit points, so that in our case we obtain

$$T = \exp\left\{-\frac{2}{\hbar}\int_0^{x_0} (2m[V(x) - E])^{1/2}\,\mathrm{d}x\right\}. \tag{8.14}$$

If the barrier has the triangular form $W - Fx$ shown in Fig. 8.4, then the transmission coefficient of an electron at the top of the reservoir (at $E = E_F$) is

$$T(E_F) = \exp\left\{-\frac{2}{\hbar}\int_0^{W/F} [2m(W - Fx)]^{1/2}\,\mathrm{d}x = \exp\left\{-4(2m)^{1/2}\frac{W^{3/2}}{3\hbar F}\right\}\right.. \tag{8.15}$$

The approximation we have used here is a simplified form of the Wentzel–Kramers–Brillouin (WKB) method that is widely used in quantum theory. It is unsuitable for applications to thin barriers and when a strong external field is involved. Then the wave function must vary too rapidly and eqn (8.11) can no longer be justified.

Having found the transmission coefficient, we can evaluate the tunnelling current density j_t at E_F:

$$j_t = -envT, \tag{8.16}$$

where v is the electron velocity at the top of the free-electron reservoir at E_F, T is from eqn (8.15) and n is the electron density. This result is valid only at low temperature, because we have ignored the thermionic emission contribution.

The general prescription for T in eqn (8.14) is easy enough to apply to

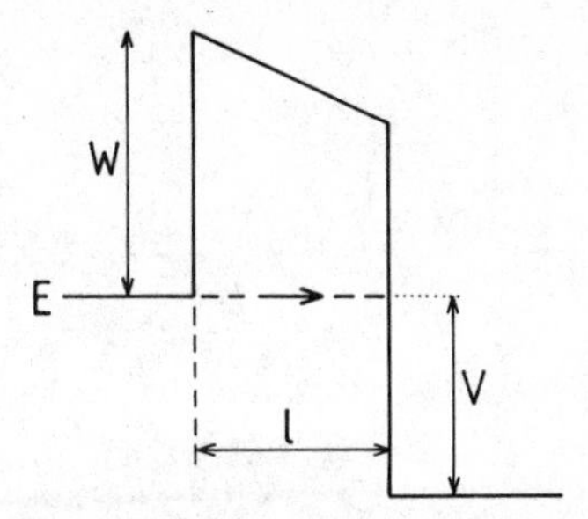

Fig. 8.5. A more general form of a repulsive barrier. The transmission coefficient is given by the Nordheim–Fowler formula, eqn (8.18).

other potential curves. In many cases, a solution can also be found by carrying out the integration numerically.

Some insight into the nature of the approximation underlying the WKB method can be obtained when the expression for T is applied to the case of a particle at $E=0$ facing the rectangular barrier of constant height V and width $2a$ shown in Fig. 8.1. The result is

$$T=\exp[-4a(2mV/\hbar^2)^{1/2}]. \quad (8.17)$$

As expected, this result is different from the exact answer obtained in Section 8.1.

An important barrier form for which an analytic expression for T is available is shown in Fig. 8.5. It is a more general step-like biased (V) barrier, and the corresponding result for T of an electron at energy E indicated in the figure is

$$T=\exp[-(2W)^{3/2}m^{1/2}l/\hbar V]. \quad (8.18)$$

This is the so called Fowler–Nordheim formula.

8.3. Tunnel junctions

The earliest application of the tunnel effect was made in tunnel diodes, also known as Zener or Esaki diodes. This device is basically a reverse-biased p–n junction. At high doping concentrations (10^{19}–10^{20} impurities cm^{-3}), the valence band of the p-type part is degenerate in energy with the conduction band of the n-type region (stage 1 on the left-hand side of Fig. 8.6). The uppermost occupied state is at the Fermi energy E_F, so that there are empty levels at the top of the valence band on the p-type side, and occupied levels in the conduction band on the n-type side. When the junction is biased (stage 2), electrons tunnel across

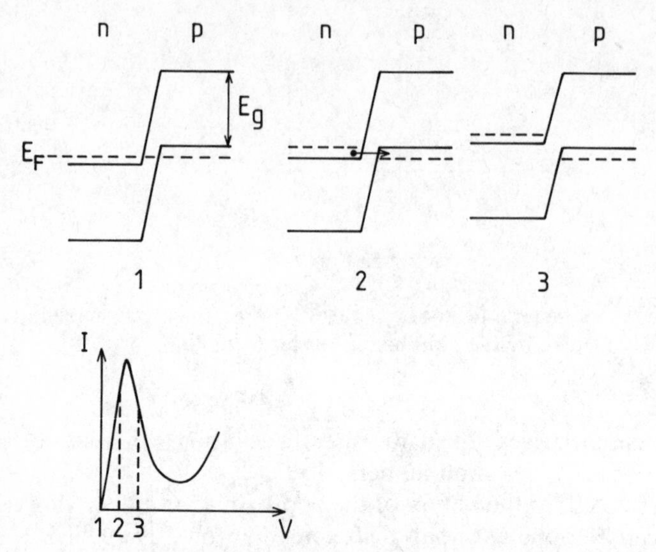

Fig. 8.6. (Top) A band diagram for a tunnel junction with three different values of the bias voltage. (Bottom) The current-voltage characteristic of a tunnel diode, with the three stages shown above indicated on the curve. E_g and E_F are the bulk band gap and the Fermi energy, respectively.

the narrow junction to fill the empty levels in the valence band on the right. The current rises with applied voltage. As the empty slots are being filled, the current passes a peak and begins to grow weaker (stage 3).. A sketch of the curent–voltage curve is given in the lower part of Fig. 8.6. The high doping concentration means that the junction is very narrow (~100 Å), as can be verified from the formalism developed in Chapter 4. Since the time of flight of the tunnelling electrons is proportional to the length of the electron path, the tunnel diode can be used as a fast switch. The most remarkable property of tunnel diodes is the negative slope of the current–voltage curve. This means that the system has negative resistance in that regime.

8.4. Velocity overshoot in hot-electron transport

The tunnel diode was the first device in which hot electron effects became important. The high doping concentration creates a large electric field at the junction in which electrons (and holes) are accelerated. In weak electric fields, the electron velocity is proportional to the applied field, and the electron mobility can be expressed in terms of a relaxation time constant τ. At high fields, electrons are accelerated so rapidly that the time it takes them to dispose of the gained kinetic energy by collisions with lattice vibrational waves (i.e. by emitting phonons) is too long to maintain thermal equilibrium. This means that the population of states is no longer given by the Boltzmann factor we have been using throughout this course. For example, at a given time and point in space, the crystal may have more electrons at an energy E than at E', where $E > E'$. Consequently, there is a range of field strengths at which the electron velocity 'overshoots' the saturation value, which is approximately equal to the maximum thermal velocity for the random motion of carriers at a given temperature. If we plot velocity as a function of distance along the channel for different values of the electric field, these curves exhibit a characteristic peak. This is illustrated in Fig. 8.7, which shows the results of detailed simulations of hot electron transport in Si and GaAs. At large

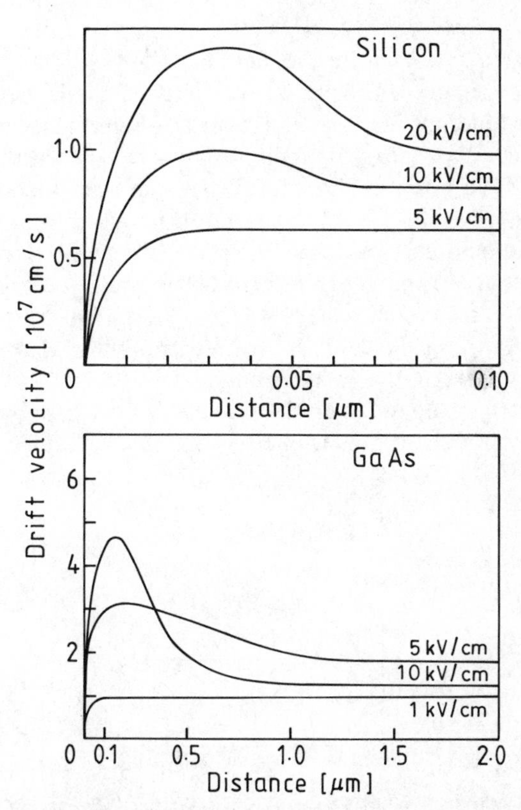

Fig. 8.7. The drift velocity in Si and GaAs in electric fields of varied strengths at different distances along the particle path in the channel.

distances, the electrons make many collisions and the average velocity drops to the saturation value. It follows that in short channels the carrier may not have enough time to make the collisions to reach the equilibrium distribution, and the operational (drift) velocity can be enhanced by a factor of 2 to 3. Such short channels can be achieved in semiconductor microstructures described in Chapter 7.

8.5. Transfer of hot electrons into secondary valleys. Negative resistance

Figure 8.7 suggests that both the absolute magnitude of the overshoot velocity and its dependence on the strength of the applied field depend on the choice of material. The enhancement is larger in GaAs because of the small effective mass at the bottom of the conduction band and consequently a low density of states. However, the enhancement achieved in GaAs by increasing the field from 5 to 10 kV is not as strong as might be expected. In fact, if even higher fields were used, the enhancement would not increase any further. The ultimate limiting factor can be identified if we recall the many-valley character of the conduction band of GaAs. We noted in Chapter 5 that there are secondary minima in the conduction band of GaAs separated by about 0.3 eV from the band edge. When the field exceeds 5 kV, the probability that electrons acquire this amount of energy and get transferred from the principal Γ valley into either the X or L secondary valleys is large. This is illustrated in Fig. 8.8. In order to accomplish the transfer, the electron must also collide with a phonon. This process changes its momentum. The effective mass at X and L is about five times larger than at the lowest Γ point. Thus, as a result of the excitation to higher energy, the electron velocity is actually

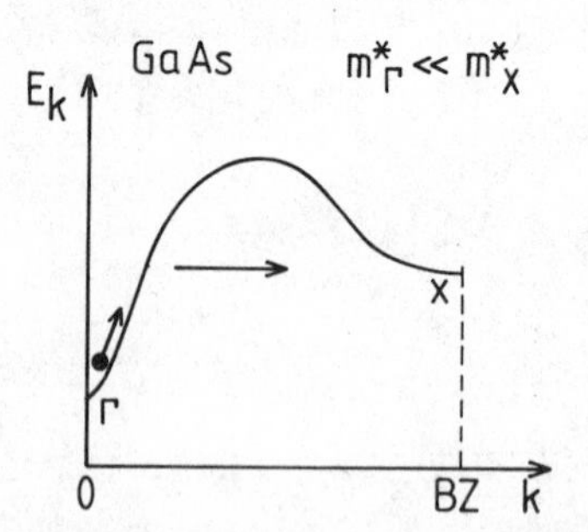

Fig. 8.8. The conduction band of GaAs between the lowest Γ and secondary X minima. An electron accelerated in an external field (solid circle) can be transferred from Γ to X as indicated. The electron effective mass at X is much larger, so that the gain in energy is accompanied by lowering of velocity.

decreased. The significance of this limitation can be diminished in very short channels. This is because the electrons may then reach the end of channel before they have had a chance to collide with a phonon of suitable momentum and accomplish the transfer.

It is worth commenting that according to this account of high-field transport in GaAs, the crystal behaves as a device with negative resistance. A number of applications have been implemented by exploiting this effect (the so-called Gunn diodes or oscillators).

8.6. Impact ionization. Avalanche current multiplication

Energetic (hot) carriers generated by high electric fields may dissipate their energy in a process called *impact ionization*. For example, an energetic electron hits an atom and loses its energy by exciting an electron from the valence band across the forbidden gap. After the collision, we have two electrons at the bottom of the conduction band and one hole at the top of the valence band. This is a process in which the cooling of energetic electrons doubles the electron population in each impact-ionization event. In bulk semiconductors, this current multiplication is limited by the electron lifetime, i.e. by the fact that there is a large probability for the electrons excited across the gap to recombine with holes. Heterojunction microstructures can be used to maximize the current multiplication. This is illustrated in Fig. 8.9.

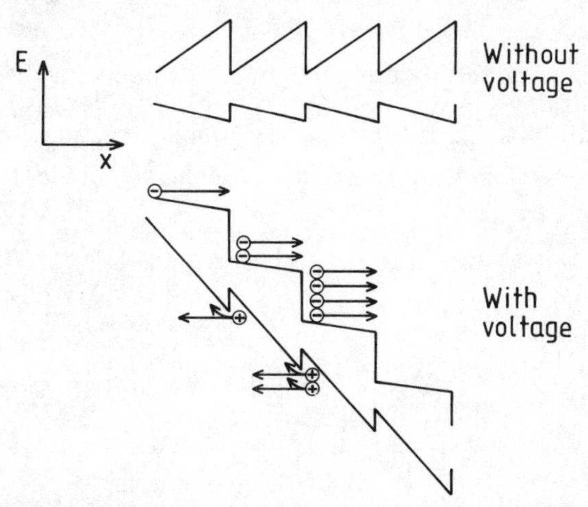

Fig. 8.9. (Top) A band diagram of a sawtooth superlattice, considered in Chapter 7. (Bottom) The band diagram with an applied electric field, indicating the avalanche multiplication of the current of electrons injected from the left. The directions of the electron and hole motions are shown by arrows.

In the upper part of Fig. 8.9, the band diagram (variation of the conduction and valence band edges) is a sawtooth superlattice such as that discussed in Chapter 7. The lower part shows the band diagram of the superlattice in an external electric field. Injected electrons gain energy in the field and impact-ionize at the interfaces. At every interface, the electron population is doubled. Since the barrier confining holes is small and that confining electrons is large, the superlattice structure makes it easy for holes to escape from the interface before a recombination event can take place. Such 'avalanche' current multiplication has been used to make efficient photodetectors.

8.7. Breakdown voltage

In most instances, the effects associated with large electric fields are unwanted phenomena. Such effects are particularly important at interfaces and p–n junctions. Consequently, when we have a junction biased by an applied voltage V, we desire to identify the conditions under which the junction should function properly. To understand the relationship between the applied voltage and the other parameters determining the junction properties, we must turn to the expression for the maximum electric field derived in Chapter 4. Assuming that the sample is only moderately doped, we can write for the n-type region, from eqns (4.23) and (4.26),

$$|\mathscr{E}_{\max}| = [2eN_d(\phi + |V|)/\epsilon_0\epsilon]^{1/2} \tag{8.19}$$

where N_d is the concentration of (fully ionized) donors and the resistivity of the sample in the n-type region is $\rho = (N_d e\mu_e)^{-1}$; μ_e is the electron mobility. As we increase the reverse-bias voltage, the barrier potential ϕ holds the leakage current down to a negligible value (tunnelling is

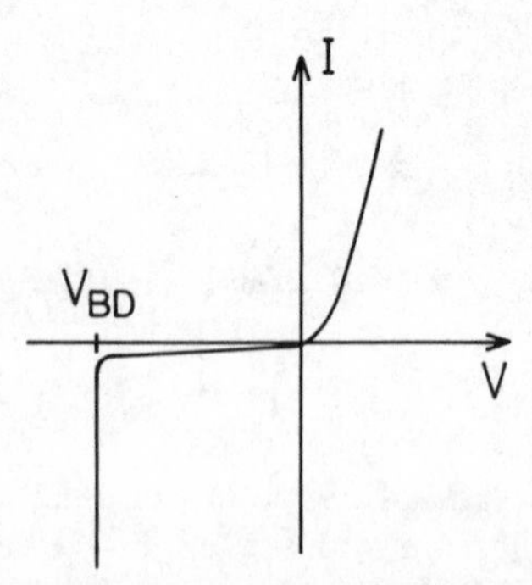

Fig. 8.10. The current–voltage characteristic of a p–n junction diode indicating the breakdown voltage V_{BD}.

significant for very thin junctions and since it is not included in eqn (8.19) it must be estimated separately). The breakdown occurs when the applied voltage is much larger than ϕ. Hence, taking $\phi \ll V$, we can estimate the breakdown value of the applied voltage V from eqn (8.19) as

$$V_{\mathrm{BD}} \simeq \rho\mu_e\epsilon_0\epsilon\,|\mathscr{E}_{\max}|^2/2. \tag{8.20}$$

It follows that the breakdown voltage V_{BD} increases with increasing resistivity of the material. The position of V_{BD} in the current (I) versus voltage (V) characteristic of a p–n junction diode is illustrated in Fig. 8.10.

8.8. Poole–Frenkel effect

The carrier mobility in a semiconductor decreases with the frequency of collisions the carrier makes with lattice waves and impurities. Collisions with impurities may lead to the carrier being captured. This means that, for example, an electron settles down into a level introduced into the forbidden gap by an impurity atom. Although it may later be excited back into the conduction band by an external field or thermally, during the time between capture and re-emission the carrier does not contribute to the current. The probability that an electron is captured into an impurity state is given approximately by the area πa^2 associated with the radius a of the orbit of the state in question. The reason for this is the same as if we were trying to assess the chance that a randomly moving very small ball hits a target of a certain cross section. For the electron to be re-emitted, it must overcome the influence of the attractive potential responsible for binding. Since electron or hole transport is inconceivable without an applied electric field, we must ask how this field affects the trapping and emission at impurities, i.e. the processes that determine the sample mobility.

The Coulomb potential V, representing a donor impurity, is shown in Fig. 8.11a. This potential is modified by the external field F, applied along the z axis, as indicated in Fig. 8.11b. The total potential energy is then

$$V(z) = \frac{-e^2}{4\pi\epsilon_0\epsilon z} - eFz. \tag{8.21}$$

The function $V(z)$ has a maximum when $\mathrm{d}V/\mathrm{d}z = 0$. This occurs at a point $z_{\max}$ from the impurity site at $z = 0$. The height of the potential barrier δE_B measured with respect to $V = 0$ is

$$|\delta E_B| = V(z_{\max}) = (e^3F/\pi\epsilon_0\epsilon)^{1/2}; \qquad z_{\max} = (e/4\pi\epsilon_0\epsilon F)^{1/2}. \tag{8.22}$$

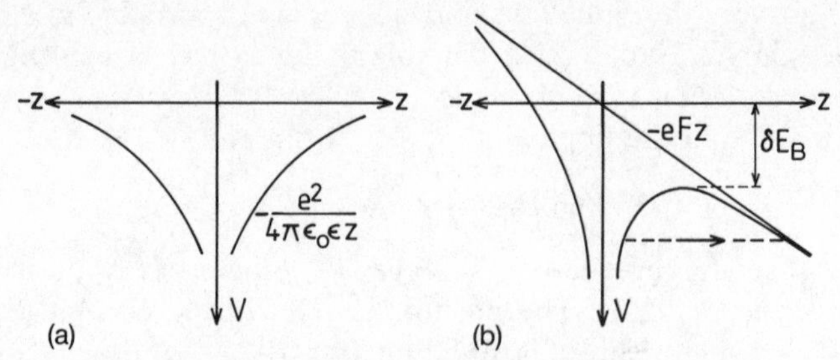

Fig. 8.11. (a) The attractive Coulomb potential energy associated with a donor impurity in a semiconductor; ϵ is the dielectric constant. (b) The potential energy in the presence of an external electric field F. δE_B indicates the lowering of the barrier seen by an electron trapped at the donor. The probability that the electron can escape by tunnelling is greatly enhanced. The tunnelling path is indicated by the arrow. A similar enhancement affects electron capture.

If, in the re-emission event, an electron is thermally excited into the conduction band, the applied field is responsible for a lowering of the barrier by δE_B and the probability of re-emission is enhanced by the factor $\exp(\delta E_B/k_BT)$. This is known as the Poole–Frenkel effect.

At low temperatures, the thermal emission process is weak. However, the electron can escape from the trap by tunnelling, which becomes a dominant process at low temperatures. It is clear that the barrier lowering due to the applied field also enhances the tunnelling probability.

If the impurity is positioned in the vicinity of a heterojunction, the potential will be altered further because of the effect of the confining barrier. The potential contributed by the heterostructure can be taken from Chapters 6 or 7 and simply added to the potential in Fig. 8.11. The tunnelling probability can be evaluated from the WKB formulae developed earlier in this chapter.

8.9. Mobility enhancement in modulation-doped structures

Semiconductor microstructures offer a very simple and effective means of increasing the speed of electronic devices. The reduction in size—given by the very nature of such structures—is the most obvious helpful factor, since the net electron transit time can be shortened simply by the reduction of the size of the device. That in itself provides a strong incentive for making electronic devices smaller and smaller. However, the size effect and its role in altering the electronic structure can be exploited directly to enhance the carrier mobility μ.

To understand how such an enhancement can come about, we must

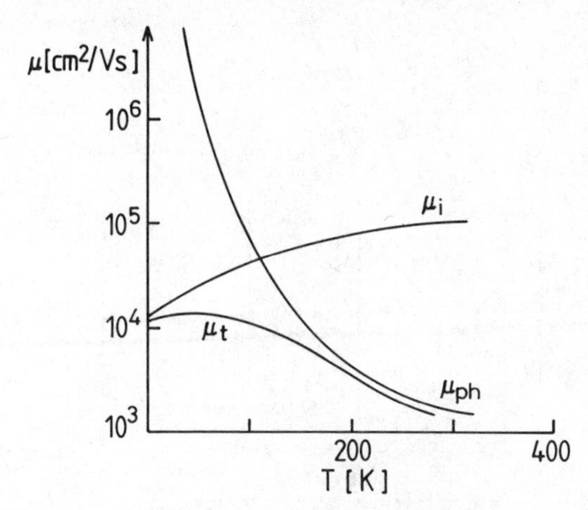

Fig. 8.12. The electron mobility μ as a function of temperature; μ_i and μ_{ph} are the impurity and the phonon contributions, respectively; μ_t is the total magnitude of the mobility.

return to the two most important processes determining the magnitude of μ, discussed in Chapter 4.

At high temperatures, the lattice vibrations are the main obstacle for electrons. The contribution to the electron mobility due to collisions with vibrational waves (phonons) is shown in Fig. 8.12 (μ_{ph}). For example, at room temperature (about 300 K) the total electron mobility is almost entirely given by the electron–phonon interactions. However, at low temperatures, the lattice waves are suppressed and the contribution to mobility μ_i due to collisions with impurities dominates. The vibrational properties of semiconductor crystals do not vary greatly from material to material and neither does the electron–phonon interaction. There is, therefore, very little hope that the mobility could be significantly improved in the high-temperature range.

In order to achieve an improvement at lower temperatures, we must find ways of preventing electrons from colliding with impurity atoms. Unfortunately, in a conventional transistor, the presence of impurities is unavoidable because they are the very dopants supplying the carriers of the electrical current whose mobility we want to be enhanced. Ideally, we would like to take the carriers away from dopants and keep them in a channel that is spatially separated from the doped region, and in which the carriers are free to move without the unwanted collisions. Heterojunction microstructures can do just that.

Take, for example, the GaAs–$Ga_{0.7}Al_{0.3}As$ quantum well structure whose conduction band diagram is reproduced in Fig. 8.13. This structure is prepared by growing layers one by one upon a semi-insulating GaAs

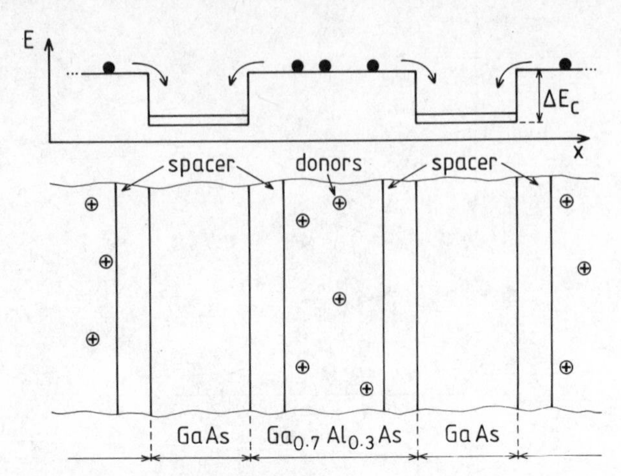

Fig. 8.13. (Top) The conduction band diagram of a quantum well structure. ΔE_c is the conduction band offset. (Bottom) The spatial arrangement of the GaAs and the alloy layers in a modulation-doped structure. Circles indicate the dopants in the alloy. The undoped alloy region is called the spacer. GaAs is high-purity undoped material.

substrate. Therefore, doping can be done separately for each layer. To reduce the possibility of in-diffusion of impurities during growth, only the middle part of the barrier material is doped. This creates spacer layers of undoped alloy separating the doped alloy from the high-purity (undoped) GaAs. Because of the band line-up discussed in Chapter 7, electrons released from impurities located in the barrier material move around (diffuse) until they eventually land at the bottom of a GaAs well. Once there, they are confined by the barrier ΔE_c and forced to remain there.

If we now apply an electric field in the direction parallel to the interface, the electron (or hole) motion in the well is free of collisions with impurities. The spacer regions in the alloy in fact also ensure that the effect of the long-range tail of the Coulomb potential, (originating from the ionized dopants in the barrier) upon the electron motion in the GaAs channel is negligible.

To take advantage of the impurity-free channel, we must cool the structure to liquid-nitrogen temperature. The electron mobility in this so-called modulation-doped structure is very much larger than the mobility in bulk GaAs. Typical experimental results for a modulation-doped structure are shown in Fig. 8.14 and compared with those obtained for bulk GaAs.

The quantum well structure pictured in Fig. 8.13 is particularly suitable for the presentation of the principle of modulation doping, but it is by no means the only arrangement whereby the principle can be implemented.

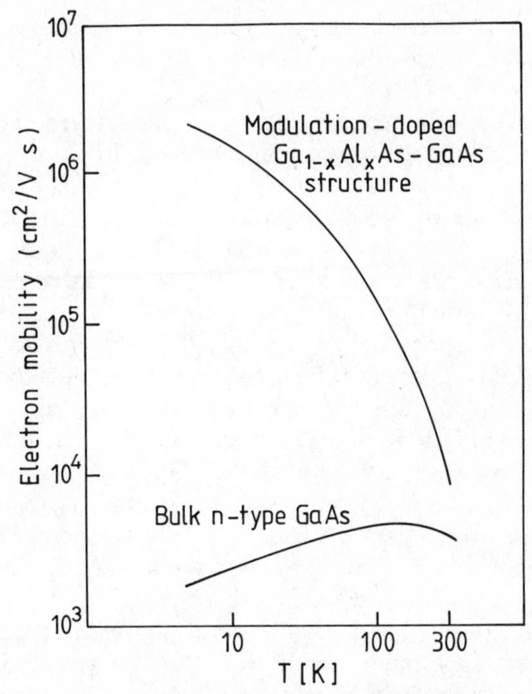

Fig. 8.14. The electron mobility as a function of temperature, in a modulation-doped structure and in bulk n-type GaAs.

The device structures employed to build high-speed transistors will be discussed in Chapter 11.

8.10. Superconductivity

Superconductivity was discovered by Onnes in 1911, soon after liquid helium became available to make low-temperature transport experiments possible. Onnes thought that the resistance of a metal should be very low at such temperatures. He found that the resistance of mercury was indeed effectively zero up to about 4.2 K, and then rose abruptly (within a transition width of less than one-hundredth of a degree) to about 0.1 ohm. He chose mercury because it was one of the few high-purity metals available to him. However, it soon became apparent that there was no significant connection between the key features of his data and the sample purity. He could reproduce his results even when substantial amounts of impurity were added. The explanation of this behaviour was provided by Bardeen, Cooper and Schriefer (BCS) in 1957 (the same

Bardeen who is credited with the discovery of the transistor!). It lies in a completely different view of the origin of electrical current in the superconducting state.

In Chapter 4, we outlined the essence of electron transport in metals. According to this model, in a simple metal electrons at the top of the free-electron reservoir at E_F are free to jump into higher empty states in response to an arbitrarily small excitation. However, in certain materials, the electron wave functions describing the states in which a conducting electron finds itself are more complicated. A conducting electron may interact strongly with the lattice and deform it. We can visualize this deformation as a response of the system of atoms forming the lattice to the presence of the electron, in the same way as we view the polarization of a lattice by a propagating light wave discussed in Chapter 9. A second electron then comes along and takes advantage of this deformation, in that the crystal potential in the vicinity of the first electron appears to it to be more attractive, and it minimizes its energy by remaining in the region of this deformation. We can say that the second electron interacts with the first—forms an electron pair—via the lattice deformation field. It is important to see this as a dynamic process.

Another way to put it is to say that the first electron excites lattice waves (phonons) as it moves on in the crystal. The second electron is then attracted into this region because it can lower its energy (and the total energy of the system), so that we end up with a pair of electrons travelling together. For this coupling to occur, the electron–lattice interaction must be strong enough to overcome the repulsive electrostatic interaction between the two electrons forming the pair. The strength of this coupling or the energy θ needed to dissociate the electron pair, is called the energy gap separating the superconducting state of these electrons from the 'normal' one. Consequently, if we increase the temperature above some critical value T_c, the thermal energy $k_B T_c$ added to each electron in the pair destroys the superconducting state. In the literature on superconductivity, it is customary to quote the value of the so-called gap parameter $\Delta = \theta/2$. For instance, Δ for Ga, Sn, and Hg at 0 K is 0.17, 0.57 and 0.82 meV, respectively. Above the critical temperature, the resistivity of the sample is finite and increases with increasing temperature as in an ordinary conductor.

The BCS theory predicts an interesting paradox; the stronger the electron interaction with the lattice, the more likely it is that the metal will be a superconductor when cooled. (In the normal regime of conduction, a stronger electron–phonon interaction means more efficient electron collisions with the lattice and consequently higher resistivity.) This result also tells us that if we want a superconductor with higher T_c, we must look for it among materials with higher resistivity! In fact, the

BCS theory predicts that the critical temperature increases exponentially with increasing electron–phonon interaction, a parameter that depends on the electronic (band structure) as well as elastic parameters of the material in question. The value of T_c for simple metals like Hg is a few kelvins. However, J. D. Bednorz and K. A. Müller discovered that some oxides with K_2NiF_4-type tetragonal crystal structure, e.g. $(SrBa)_xLa_{2-x}CuO_{4-y}$, have T_c in the region of liquid nitrogen temperature, i.e. around 90 K. Related oxide compounds with other metal elements from the lanthanide series exhibit even higher critical temperatures.

When two normal metals are separated by an insulating layer, the insulator acts as a barrier to the flow of electrons from one metal to the other. If the barrier is sufficiently thin (about 20 Å or less), there is a significant probability that an electron can tunnel through the barrier. At low applied electric field, the current–voltage relation of the sandwich obeys Ohm's law, as shown in Fig. 8.15a. If one of the metals is a

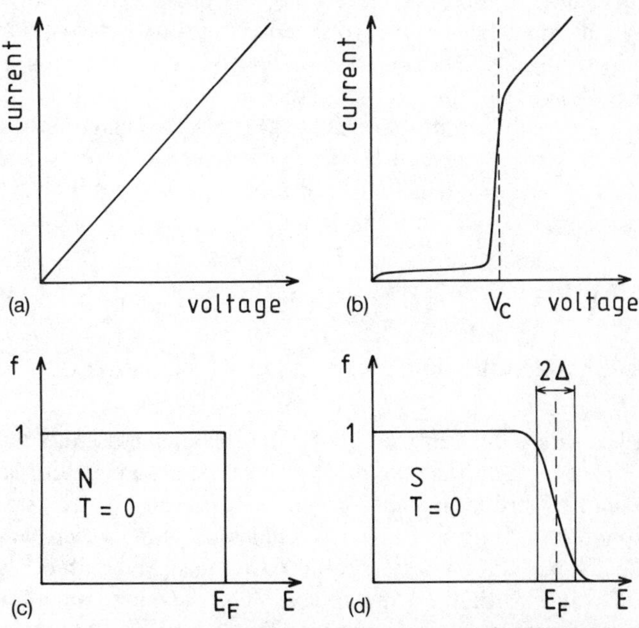

Fig. 8.15. (a) Linear current–voltage relation for tunnelling between two normal metals separated by a thin insulating layer. (b) Current–voltage relation when one of the metals is a superconductor. The threshold voltage V_c for ohmic behaviour is equal to Δ/e, where Δ is the gap parameter. (c) The normalized density of occupied states in a normal metal (N) and (d) in a superconductor (S), showing the position of the superconductor gap 2Δ and the free-electron Fermi energy E_F.

superconductor, the relation changes to that in Fig. 8.15b. At 0 K, the ohmic behaviour is established when the applied voltage V exceeds the magnitude of the gap parameter Δ, i.e. $eV = \Delta(=\theta/2)$. This allows us to visualize θ simply as a gap in the energy spectrum, in the manner familiar from Chapter 2. The continuous distribution of states (Fig. 8.15c) that is characteristic of a normal metal (electron gas) near the Fermi energy is replaced by a discontinuous spectrum with a threshold voltage $eV = \Delta$ needed to take an electron from the normal metal to the superconductor (Fig. 8.15d). At temperatures different from zero, there is a small current even at lower voltages, because some electrons in the superconductor are thermally excited across the energy gap θ.

Whereas liquid-helium temperatures are difficult and expensive to work with, liquid-nirogen temperatures are not and are commonly used in device applications. Indeed, the principle of modulation doping discussed in Section 8.9 can only be taken advantage of if the semiconductor microstructure is cooled to that temperature. The high-temperature superconductors therefore offer the possibility of providing highly efficient material for interconnections and other components used in microelectronics. However, the materials exhibiting this effect are really ceramics unsuitable for manufacture of ultrasmall structures. Although it is possible to prepare single crystals of such compounds, their lattices are full of imperfections and macroscopic defects and do not support strong current density. They are therefore unlikely to replace semiconductors as raw material for transistors in highly integrated circuit systems.

8.11. Josephson tunnelling

We have seen that the zero-resistivity state found in some solids can be explained if we assume that the basic unit of electrical current is a pair of weakly bound electrons. Because of the very nature of its existence, such an electron pair is not hindered by collisions with lattice waves. This means that over a certain distance the pairs must be treated as coherent waves, i.e. waves with the same phase. This is quite analogous to the behaviour of photons in a laser beam. The coherence length is of order 500 Å and scales with the mean free path characteristic of the normal metallic behaviour of the sample.

We can assume that the common phase of the superconducting electron pairs varies with time t in an analogous way to the phase of an ordinary electron wave function. From elementary quantum mechanics, we know that the time dependence of the wave function ψ of an electron of energy E is

$$\psi \sim \exp(\mathrm{i}Et/\hbar). \tag{8.23}$$

Because a superconductor has zero resistance, it cannot sustain potential differences. Consequently, E in eqn (8.23) is a constant. More generally, we can write the wave function of our pair as

$$\psi \sim \exp[\mathrm{i}\phi(t)], \tag{8.24}$$

where $\delta\phi/\delta t = E/\hbar = \text{constant}$. This result has interesting consequences if we consider two superconductor layers separated by a thin insulator of width $2a$ (Fig. 8.16). It is clearly possible to apply a voltage to such a system. The tunnelling superconducting current of electron pairs must then reflect the phase difference of the wave functions in the two conducting layers.

Let us imagine the superconductor wave functions in our junction sandwich structure decaying from the superconductor into the insulator on both sides. The two superconductors are made of identical material and the electron concentration in them must be the same. The wave function amplitude representing the current must therefore be also the same on both sides of the insulator. The phases are in general different and we shall label them, as in eqn (8.24), ϕ_1 and ϕ_2. The total wave function in the insulating layer is a superposition of the two decaying contributions that are pictured in Fig. 8.16:

$$U(x) = C\{\exp(\mathrm{i}\phi_1)\exp[-\beta(x+a)] + \exp(\mathrm{i}\phi_2)\exp[\beta(x-a)]\}, \tag{8.25}$$

C is a normalization constant.

The current density is $j = nqv$. We shall replace the classical electron number density n by the quantum mechanical expression (probability

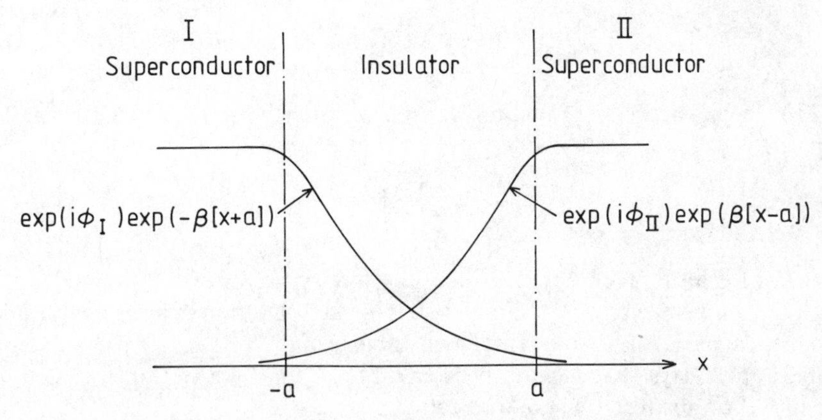

Fig. 8.16. A Josephson junction (described in the text).

density) $U(x)U^*(x)$. We assume that the mean value of the probability density is n, so that C is porportional to n. The velocity is $v = p/m$, where p is the linear momentum. In quantum theory, p is an operator expression $\hat{p} = -i\hbar(\delta/\delta x)$. The quantum expression for j is therefore

$$j = -\frac{i\hbar q}{2m}\left(U^*\frac{\partial U}{\partial x} - U\frac{\delta U^*}{\delta x}\right). \tag{8.26}$$

Substituting from eqn (8.25) into eqn (8.26), and carrying out the differentiations with respect to x, leads to

$$j = j_0 \sin(\phi_2 - \phi_1), \tag{8.27}$$

where j_0 is a constant characteristic of the junction. This is the basic equation for the junction derived by Josephson. If there is a potential difference V across the junction, then the difference in E of eqn (8.23) between the two superconductors is $E_2 - E_1 = qV$. Therefore, we have from eqn (8.24)

$$\phi_2 - \phi_1 = \frac{qVt}{\hbar} + \delta\phi. \tag{8.28}$$

Substituting into eqn (8.27) we obtain for the junction current density

$$j = j_0 \sin(qVt/\hbar + \delta\phi). \tag{8.29}$$

This equation predicts that, in the absence of an applied voltage, j is constant and can be adjusted by $\delta\phi$. When V is non-zero, there is an oscillatory current with angular frequency $qV/\hbar$. If the junction is irradiated with a beam of frequency ω, an additional voltage appears across the junction of magnitude $W\cos(\omega t)$, and the result in eqn (8.29) is modified so that it now gives

$$j = j_0 \sin\left\{\frac{q}{\hbar}\left[Vt + \frac{W}{\omega}\sin(\omega t)\right] + \delta\phi\right\}. \tag{8.30}$$

This means that the junction carries frequency-modulated current. In particular, a d.c. (zero-frequency) current is obtained if

$$V = \frac{l\hbar\omega}{q}, \tag{8.31}$$

where l is an integer. Since $q = 2e$, the theory predicts well-defined steps in the current–voltage characteristic of the junction whose amplitude depends only on fundamental constants e and $\hbar$. The relation in eqn (8.31) also says that a photon of energy $\hbar\omega$ that is a multiple of $2eV$ is emitted or absorbed when an electron pair crosses the barrier.

This account of the current–voltage characteristics of a superconducting junction shows that such a junction can be used as a fast switch. Although this device has not been commercially exploited, a computer based on the Josephson junction principle outlined here has been built and its functioning has been demonstrated. Unfortunately, the project was eventually abandoned because of difficulties with the low-temperature equipment and the fragility of the ultrathin tunnel junctions.

Problems

8.1. Consider a $Ga_{0.7}Al_{0.3}As$ alloy barrier layer 50 Å thick. Calculate the transmission coefficient T for electrons at $E = 50$ meV above the conduction band edge of GaAs, at zero temperature.

8.2. Consider a $Ga_{0.7}Al_{0.3}As$ barrier 40 Å thick and an external electric field of 500 kV cm^{-1}. Estimate the transmission coefficient for an electron of energy 50 meV above the conduction band edge of GaAs.

8.3. Determine the peak value of the electric field of a silicon p–n junction reverse-biased by 6 V (i.e. near the breakdown voltage). Take $N_d = 2 \times 10^{18}$ cm^{-3} and $N_a = 5 \times 10^{19}$ cm^{-3}.

8.4. Estimate the magnitude of an applied electric field such that the barrier lowering in the Poole–Frenkel effect equals the binding energy of a donor in GaAs.

8.5. The current in a 10 Å thick Josephson junction oscillates with frequency 483.6 MHz. Find the magnitude of the applied electric field in V cm^{-1} (i.e. d.c. voltage).

8.6. Estimate the electron kinetic energy needed for an impact ionization event to exist in InP, GaAs, and ZnSe. How long would it take for the hot electron to relax to the bottom of the conduction band of GaAs if it disposed of its kinetic energy by interacting with the lattice?

8.7. Repeat Problem 8.4 for a donor in silicon and compare the magnitude of the field in the two cases.

8.8. Calculate the tunnelling current density for Problem 8.2 and compare it with the contribution to current due to thermionic emission at room temperature.

8.9. Consider a p–n junction in silicon at 300 K, with 10^{14} donors cm^{-3} and 5×10^{19} acceptors cm^{-3}. Estimate the drift and diffusion currents (the diffusion constant is $D = 12.5$ cm^2 s^{-1}).

9

Optical properties of microstructures

9.1. Propagation of electromagnetic waves in dielectrics. Linear response

In our discussion of the electronic structure of quantum wells and superlattices, we made frequent references to the relationship between the magnitude of the forbidden gap and the optical properties of such systems. We were concerned with the problem of 'engineering' structures in which emission or absorption occur at the desired value of energy of the light wave. For example, such an emission process corresponds to a jump of an electron from the bottom of the conduction band to an empty slot (hole) at the top of the valence band. Hence, this process can be characterized by the energy or wavelength of the emitted photons, by the lifetime τ of the electrons in the conduction band, and by the strength of the optical transition (absorption coefficient α), which is a measure of the probability with which such transitions take place. In Chapter 12 we shall use these parameters to describe the design and functioning of heterojunction lasers and light-emitting diodes. Since τ, α, the light wave wavelength λ, and the energy are treated as constants, the underlying assumption is that these parameters are independent of the number of electrons making transitions or the number of photons in the light beam. We can also describe this state of affairs by saying that the refractive index n of the system is a constant, or that the response of the system to a propagating electromagnetic wave is independent of the intensity of the light beam. By *system* we mean the solid in which the optical process takes place, and whose properties we can express in terms of the electronic band structure. By *response* we mean the ability of the valence electrons sitting in the quantum states forming the valence band to redistribute their charge density slightly when they 'see' the beam of light waves. We say that electrons are polarized by the beam.

Essentially the same happens when a free atom, say hydrogen, is exposed to a beam of light. The electron orbital around the proton, which in the absence of the external field is a sphere, is then elongated along the axis of the electric field vector. We say that the field induces a small dipole component into the electron charge distribution. Since electrons in free atoms are very tightly bound to the nucleus, and reside in orbits of a very small radius, the changes in the electron charge distribution induced by the light beam are relatively insignificant. However, in solids, electrons are shared by atoms, and their wave functions are not so well localized. They are, therefore, more likely to be affected (polarized) by external fields. This may in turn influence the propagation of light waves in the solid.

The qualitative features of the response of a polarizable system (e.g. electrons bound to atoms forming a solid) to an external field of given frequency ω can be recovered in terms of a simple classical model in which an electron is visualized as a classical particle of mass m oscillating in a one-dimensional harmonic potential $\frac{1}{2}\beta x^2$. If the external field is a harmonic function $\mathscr{E} = \mathscr{E}_0 \exp(\mathrm{i}\omega t)$, then the equation of motion describing the response of our electron to $\mathscr{E}$ is

$$m\ddot{x} + \sigma\dot{x} + \beta x = -e\mathscr{E}_0 \exp(\mathrm{i}\omega t) \tag{9.1}$$

where β is the classical restoring force and σ is the damping coefficient. We can define the resonance frequency $\omega_0 = (\beta/m)^{1/2}$ and assume that the solution $x(t)$ of eqn (9.1) is a harmonic function with amplitude x_0, i.e.

$$x(t) = \mathrm{Re}[x_0(\omega)\exp(\mathrm{i}\omega t)]. \tag{9.2}$$

Re means that we take the real part of the expression. Substitution into eqn (9.1) gives

$$(\omega_0^2 - \omega^2)x_0 + \mathrm{i}\omega\sigma x_0 = -e\mathscr{E}_0/m, \tag{9.3}$$

and the amplitude x_0 is

$$x_0(\omega) = -\frac{e\mathscr{E}_0}{m}\frac{1}{\omega_0^2 - \omega^2 + \mathrm{i}\omega\sigma}. \tag{9.4}$$

We say that the electron is polarized by the external field, which induces a dipole $-ex(t)$. If there are N electrons in unit volume, the total polarization is

$$P(\omega) = -Nex_0. \tag{9.5}$$

The electronic susceptibility $\chi(\omega)$ is defined as the ratio of the complex amplitude of the induced polarization to the amplitude of the external field multiplied by ϵ_0,

$$\chi(\omega) = \frac{P(\omega)}{\epsilon_0\mathscr{E}_0}. \tag{9.6}$$

Since $P(\omega)$ is a complex function, $\chi(\omega)$ must have real and imaginary parts $\chi'(\omega)$ and $\chi''(\omega)$, respectively. We can express the induced dipole moment $p(t)$ in terms of χ by substituting from eqns (9.2)–(9.6):

$$p(t) = \mathrm{Re}[\epsilon_0 \chi(\omega) \mathscr{E}_0 \exp(\mathrm{i}\omega t)]. \tag{9.7}$$

In the classical theory of electromagnetism, the propagation of light waves in polarizable media—the so-called dielectrics—is described in terms of the electric displacement vector D such that

$$D = \epsilon_0 \mathscr{E} + P = \epsilon_0 \mathscr{E} + \epsilon_0 \chi \mathscr{E} = \epsilon_0 \epsilon \mathscr{E}, \tag{9.8}$$

where $\epsilon = (1 + \chi)$ is the dielectric constant.

We must now identify the changes in the propagation of an electromagnetic wave caused by the polarization process represented by the susceptibility in eqn (9.6). Therefore, we want to isolate the effect of χ from other contributions that may play a part in determining the dielectric constant. Let us, therefore, define for our present purposes an effective dielectric constant ϵ' such that

$$D = \epsilon_0 \mathscr{E} + P^0 + P = \epsilon_0 \epsilon' \mathscr{E}; \qquad \epsilon' = \epsilon(1 + \chi/\epsilon). \tag{9.9}$$

The dielectric constant ϵ accounts for contributions associated with the polarization process P^0, whereas ϵ' accounts for both, P^0 and P. Each of the polarization processes is assumed to be independent.

Let us consider a plane wave propagating in a medium characterized by ϵ'. The wave has the form

$$\mathscr{E} = \mathscr{E}_0 \exp[-\mathrm{i}(k'x - \omega t)]; \qquad k' = \omega(\mu_0 \epsilon_0 \epsilon')^{1/2}. \tag{9.10}$$

The wave vector k describing an electromagnetic wave of wavelength λ in the medium whose dielectric constant is ϵ is

$$k = 2\pi/\lambda = \omega(\mu_0 \epsilon_0 \epsilon)^{1/2}. \tag{9.11}$$

We can use the definition of ϵ' to relate k' and k:

$$k' = \omega(\mu_0 \epsilon_0 \epsilon')^{1/2} = \omega[\mu_0 \epsilon_0 \epsilon(1 + \chi/\epsilon)]^{1/2} = k(1 + \chi/\epsilon)^{1/2}. \tag{9.12}$$

Assuming that away from resonance $|\chi/\epsilon| \ll 1$, eqn (9.12) gives

$$k' \simeq k(1 + \chi/2\epsilon). \tag{9.13}$$

Let us express the susceptibility χ as a complex function

$$\chi(\omega) = \chi'(\omega) - \mathrm{i}\chi''(\omega) \tag{9.14}$$

and substitute into the expression for k'. We obtain

$$k' = k\left[1 + \frac{\chi'(\omega)}{2n^2}\right] - \mathrm{i}k\frac{\chi''(\omega)}{2n^2}, \tag{9.15}$$

where we have used $n = \epsilon^{1/2}$ for the refractive index of our dielectric.

If we use k' in the functional form for the propagating plane wave, we get

$$\mathscr{E}(x, t) = \mathscr{E}_0 \exp[-\mathrm{i}(k'x - \omega t)] = \mathscr{E}_0 \exp[\mathrm{i}\omega t - \mathrm{i}(k + \Delta k)x] \exp(\gamma x/2), \tag{9.16}$$

where

$$\Delta k = \frac{k\chi'}{2n^2} \tag{9.17}$$

is a phase delay per unit length that arises as a result of the polarization process of eqn (9.5). Hence it is possible to express in terms of χ' the difference in phase of the incident and outgoing fields corresponding to a change in the phase velocity of the incident wave.

The amplitude of the propagating wave decreases exponentially with distance x, with the characteristic rate of

$$\gamma = \frac{-k\chi''}{n^2}. \tag{9.18}$$

The average power w per unit volume expended by the field on inducing the electric dipole p is defined as

$$w = \left[\mathscr{E}(t)\frac{\mathrm{d}p(t)}{\mathrm{d}t}\right]_{\mathrm{av}} = \tfrac{1}{2}\,\mathrm{Re}[\mathscr{E}_0 \mathrm{i}\omega P]. \tag{9.19}$$

Using our expression for complex susceptibility and eqn (9.6), eqn (9.19) can be written as

$$w = \tfrac{1}{2}\,\omega\epsilon_0\chi''|\epsilon_0|^2. \tag{9.20}$$

The energy dissipation rate w is linked to the attenuation of the wave amplitude, which we express via γ in eqn (9.18). Indeed, eqn (9.16) implies that we assume that the intensity of the beam I decays as

$$I = I_0 \exp(-\gamma x), \tag{9.21}$$

so that the coefficient γ is

$$\gamma = \frac{-\mathrm{d}I/\mathrm{d}x}{I}, \tag{9.22}$$

which is the power absorbed per unit volume divided by beam intensity. Since γ is proportional to the imaginary part of susceptibility, in the above expressions we have established a link between χ'' and the attenuation of the wave via energy absorbed in the polarized system.

So far, we have considered the response functions χ at frequencies away from resonance. At resonance, when $\omega \to \omega_0$, the expression for χ suggests that the susceptibility (and ϵ) is greatly enhanced.

We know that we are really dealing with a system in which the polarizable particles (electrons) reside in levels that must be calculated from the Schrödinger equation. When absorption takes place (that is, at resonance), electrons make jumps between quantum states available to them, which in an intrinsic semiconductor structure at low temperatures means across the forbidden gap E_g. This implies that in our quantum system (e.g. a semiconductor crystal or microstructure) the resonance occurs when $\omega_0 \rightarrow E_g/\hbar$. Hence, we can still apply the classical model of polarization based on eqn (9.1) to intrinsic semiconductors, but we must remember that $\omega_0 = E_g/\hbar$.

It is clear from the above argument that the band gap plays an important part in determining the polarization properties of a semiconductor material. In a doped material, some of the valence states may be empty, or, in an n-type structure, some of the conduction band states may be occupied, and other resonant transitions are possible. In such circumstances, the frequency ω_0 may correspond to a separation of various levels at the valence or conduction band edges and must be specified according to the process in question.

The importance of the electronic band structure in determining optical constants of semiconductors is apparent even when the frequency of the applied field is much smaller than $E_g/\hbar$. Then—if we are not interested in attenuation effects—we can neglect ω and the damping constant in the denominator in eqn (9.4). Substituting for x_0 into the expression for P, we obtain for χ from eqn (9.6),

$$\chi = \frac{Ne^2}{\epsilon_0 m \omega_0^2}. \tag{9.23}$$

Using the definition $\epsilon = 1 + \chi$ and $\hbar\omega_0 = E_g = E_c - E_v$, where E_c and E_v are the energies of band edge levels, we find ϵ in the form

$$\epsilon^0 = 1 + \frac{Ne^2\hbar^2}{\epsilon_0 m E_g^2} = 1 + \left(\frac{\hbar\omega_p}{E_g}\right)^2. \tag{9.24}$$

This is the so-called optical dielectric constant for insulators, referred to in Chapter 5. The experimental values of this constant for different semiconductor crystals can also be found there. Because of the conditions under which eqn (9.24)was derived, the experiment must be arranged so that the data is obtained at frequencies lying well away from $E_g/\hbar$. However, ω must be sufficiently high, since precautions must be taken to avoid resonances that may be picked up due to absorption at lattice vibrational frequencies. We can model the motion of atoms under the influence of an external electromagnetic field in exactly the same manner that we modelled electron polarization, i.e. in terms of the simple oscillator equation (9.1). The restoring force is then represented by the elastic forces between atomic nuclei in the solid, considered briefly in Chapter 4 in connection with lattice vibrational waves (phonons). Since nuclei are much heavier than electrons, their energy spectrum lies at much longer wavelengths—in the far-infrared band.

9.2. Origin of non-linear effects

We chose as the starting point of our discussion of electron polarization effects in Section 9.1 a model in which the polarized particle moves in a harmonic potential of the form $V = \frac{1}{2}\beta x^2$. Since force is given by differentiating V, our choice of V implies that the restoring force is proportional to displacement (Hooke's law). The solution of eqn (9.1) describes the motion of a harmonic oscillator. We have seen that the response of this system to an external field is a linear function of the applied field, so that the susceptibility and the dielectric constant are independent of the field intensity. We argued that the classical picture can be retained to describe the essence of polarization of electrons localized at an atomic nucleus or in a solid, provided we replace the characteristic oscillator frequency by the separation in energy between the uppermost occupied and the lowest empty quantum levels divided by $\hbar$.

However, in general, the potential seen by an electron in a solid may take on a more complicated form. To account for that in our theory, we must assume that the harmonic potential of eqn (9.1) is only the first term in an expansion of V in powers of x around the origin ($x = 0$), i.e. that

$$V(x) = ax^2 + bx^3 + cx^4 + \cdots . \tag{9.25}$$

In a crystal with inversion symmetry, the potential must be symmetric and only even powers survive in eqn (9.25). (In the expansion (9.25), the term linear in x is not included. The expansion coefficient in this term must always be zero since otherwise the crystal structure would be unstable with respect to an infinitesimal displacement.)

In order to examine the role of the higher-order terms in V, let us retain for the sake of simplicity only the cubic contribution and neglect all the remaining higher terms. Let us write a new equation of motion:

$$m\ddot{x} + \sigma\dot{x} + \beta x + \delta x^2 = -\tfrac{1}{2}e\mathscr{E}_0[\exp(\mathrm{i}\omega t) + \exp(-\mathrm{i}\omega t)]. \tag{9.26}$$

As before, we assume that the external field is characterized by frequency ω. The restoring force in eqn (9.26) is a quadratic function of x, hence the labels *non-linear oscillator* and *non-linear effects*. The new term in the potential gives rise to a component oscillating at twice the frequency of the driving field, so that we can write the solution in the form

$$x(t) = \tfrac{1}{2}[q_1 \exp(i\omega t) + q_2 \exp(i2\omega t) + \text{c.c.}]. \qquad (9.27)$$

If we substitute for x from eqn (9.27) into eqn (9.26), differentiate, and equate the coefficients associated with terms containing ωt, we recover the amplitude of the motion of our linear oscillator in eqn (9.4), namely

$$q_1 = -\frac{e\mathscr{E}_0}{m}\frac{1}{\omega_0^2 - \omega^2 + i\omega\sigma}; \qquad \omega_0 = \left(\frac{\beta}{m}\right)^{1/2}. \qquad (9.28)$$

Equating the terms with $2\omega t$, we obtain the second coefficient in eqn (9.27):

$$q_2 = -\frac{e^2\delta}{2m^2}\mathscr{E}_0^2[(\omega_0^2 - \omega^2 + i\omega\sigma)^2(\omega_0^2 - 4\omega^2 + 2i\omega\sigma)]^{-1}. \qquad (9.29)$$

The induced dipole moment associated with ω is, as in eqn (9.7),

$$p^{(1)} = \text{Re}[-eNq_1 \exp(i\omega t)] = \text{Re}[\epsilon_0\chi^{(1)}\mathscr{E}_0 \exp(i\omega t)], \qquad (9.30)$$

where $\chi^{(1)} \equiv \chi$ of eqn (9.6), and $p^{(1)} \equiv p$ eqn (9.7). The contribution associated with the 2ω component is

$$p^{(2)} = \text{Re}[-eNq_2 \exp(i2\omega t)], \qquad (9.31)$$

which can be written as

$$p^{(2)} = \text{Re}[\epsilon_0\chi^{(2)}\mathscr{E}_0^2 \exp(i2\omega t)]. \qquad (9.32)$$

There are now two contributions to polarization, one of which, $P^{\omega} = \epsilon_0\chi^{(1)}\mathscr{E}_0$, is given by the familiar susceptibility χ of eqn (9.6), which we shall call the first-order or linear susceptibility $\chi^{(1)}$. The other contribution, $P^{2\omega}$, is associated with the 2ω component in eqn (9.29) and, from eqn (9.32), can be written by analogy with eqn (9.6) as

$$P^{2\omega} = \epsilon_0\chi^{(2)}\mathscr{E}_0^2, \qquad (9.33)$$

where $\chi^{(2)}$ is the second-order susceptibility. The polarization is no longer increasing linearly with the applied field since it contains $\mathscr{E}_0^2$. This means that the dielectric constant ϵ defined in eqn (9.8), which we used to express the response of the system to the field, is now

$$\epsilon = 1 + \frac{P}{\epsilon_0\mathscr{E}_0} = 1 + \chi^{(1)} + \chi^{(2)}\mathscr{E}_0, \qquad (9.34)$$

i.e. it is a function of the applied field. It follows that the phase velocity and the refractive index must also depend on the field.

If the external field consists of two beams with frequencies ω_1 and ω_2,

$$\mathscr{E} = -\frac{e}{2}[\mathscr{E}_0^{\omega_1} \exp(i\omega_1 t) + \mathscr{E}_0^{\omega_2} \exp(i\omega_2 t) + \text{c.c.}], \qquad (9.35)$$

then the polarization induced in the medium must contain, in addition to the components with ω_1 and ω_2, contributions at the sum and difference frequencies, e.g. $\omega_1 + \omega_2$ such that

$$P^{\omega_1+\omega_2} = \epsilon_0 \chi^{(2)} \mathscr{E}_0^{\omega_1} \mathscr{E}_0^{\omega_2}. \tag{9.36}$$

We have noted that we can establish a link between the quantum-mechanical band structure description of the electronic behavior and the classical formulae developed from eqn (9.1) if we associate the resonant frequency ω_0 with the difference between the uppermost occupied and the lowest empty quantum states ($\omega_0 = E_g/\hbar$). When the energy of the photons in the applied beam equals E_g, absorption takes place and the dielectric response function increases. A similar observation can be made about the second-order (non-linear) contribution to polarization. The second-order susceptibility can be enhanced by aiming two beams with photon energies $\hbar\omega_1$ and $\hbar\omega_2$ equal to the energy differences between the lowest occupied state and two (different) empty states. Of course, the classical oscillator that we employ in eqn (9.26) to model the polarization effect in solids has only one characteristic frequency (ω_0). This is equivalent to assuming that our quantum system has only one empty state (only one possible excitation energy $\hbar\omega_0 = E_g$). We know that in a semiconductor (or indeed in any other solid) there are in fact many such higher-lying states to which electrons can be excited, and consequently there are many other transition energies at which a resonance may occur. Since the purpose of our model is merely to illustrate the key concepts, we shall make no attempt to expand it so that the multitude of possible resonances is explicitly accounted for. However, we could do so by, for example, replacing our oscillator with two oscillators, with different characteristic frequencies ω_1 and ω_2.

Let us now consider the system's response away from resonance. If there is no significant absorption, and therefore no loss of energy, no electron jumps actually take place. And yet we know that the system responds to an external field and that polarization is strong, since, in the limit $\omega \to 0$, ϵ is of order 10 in semiconductors. Equation (9.24) for ϵ suggests that the response still depends on the magnitude of the energy gaps. It is as if in this case the electrons were jumping up and down by E_g in response to the external field 'virtually', i.e., without any exchange of energy. Hence we can visualize this linear response as indicated in Fig. 9.1a. It is as if our valence electron absorbed and emitted one 'virtual' photon. We have argued that in the second- (and higher-) order process, multiple jumps are involved (e.g. $\omega_1 + \omega_2$), so that our electron absorbs and emits two (three, four, etc.) virtual photons. An example of one such higher-order process is shown in Fig. 9.1b. In a full quantum-mechanical theory, we would also have to take into account that the probability that

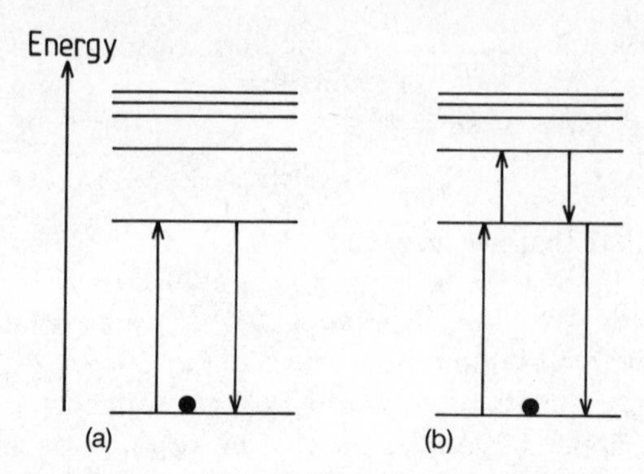

Fig. 9.1. An electron (solid circle) residing in its ground state. The lines above indicate positions of higher-lying (empty) quantum states of the system that can be obtained by solving he Schrödinger equation in the absence of any external fields. (a) An example of excitations that give rise to linear first-order susceptibility; (b) an example of a higher-order process.

a jump from one quantum state to another takes place may be less than 1, and that it depends on the details of the wave functions associated with the states in question. Having used classical equations, we have assumed that this probability is 1 (or zero; for example, we exclude the possibility that an electron jumps from one occupied state to another occupied state). In many instances, this approximation is quite realistic (as we saw, for example, in Chapter 6 when we compared our predictions of the heterojunction band offsets, which were based on our expression for the dielectric constant, with experiment).

We can now generalize the definition of polarization, along the lines indicated in the above paragraphs, to include the non-linear (higher-order) contributions. Let us define P as a Taylor series in powers of the field:

$$P = \sum_n \epsilon_0 \chi^{(n)} \mathscr{E}_0^n, \tag{9.37}$$

where $\chi^{(n)}$ is the nth order susceptibility and

$$\mathscr{E}_0^n \equiv \mathscr{E}_0(k_1, \omega_1)\mathscr{E}_0(k_2, \omega_2) \cdots \mathscr{E}_0(k_n, \omega_n). \tag{9.38}$$

Each component of the electromagnetic field is characterized by its wave vector k and frequency ω. The mixing of waves implied by the expression for the second- (third- and higher-) order contribution to P must obey conservation laws for linear momentum ($\hbar k$) and energy ($\hbar\omega$), so that not all combinations of $\omega_1, \omega_2, \ldots, \omega_n$ and $k_1, k_2, \ldots, k_n$ are likely to be allowed. As we pointed out earlier, in symmetric structures only the terms with even powers of coordinate in the potential are finite. In such materials, the lowest non-linear contribution to P is $\chi^{(3)}$. $\chi^{(2)}$ is large in strongly asymmetric crystals.

We can see from the definition of χ that the series in eqn (9.37) converges rapidly. This is because the susceptibility is inversely proportional to the powers of ω, which increase with increasing order of χ. It transpires that the higher-order terms can be increased if the systems is chosen so that the separation between the higher-lying quantum levels shown in Fig. 9.1b is small and the probability of a transition between them high (close to 1). This does not occur in crystals found in nature and, away from resonance frequencies, the second- and third-order susceptibility contributions to P are many orders of magnitude smaller than the first-order term.

9.3. Wave mixing

The existence of significant higher-order contributions to polarization gives rise to a variety of effects, all of which are generally lumped together under the label non-linear phenomena. The field concerned with applications of these phenomena in generation, propagation, and detection of light waves is known as the field of non-linear optics.

The most apparent application of non-linear effects is the possibility of converting the frequency of an optical beam up or down by mixing waves of different frequencies. For example, a large $\chi^{(2)}$ can be used for second-harmonic generation, i.e. for doubling the frequency of the incoming applied beam. A prerequisite for efficient second-harmonic generation is that coherent mixing is achieved, i.e. $k(2\omega)=2k(\omega)$, which implies $\Delta k=0$. If Δk is finite, then the constructive mixing takes place over a limited length l_c such that

$$l_c=\frac{2\pi}{\Delta k}, \tag{9.39}$$

where l_c is the *coherence length* of the system. For distances larger than l_c, interference effects become significant and the contribution to the double-frequency generated beam is small. The power of the generated second-harmonic beam is therefore proportional to l_c. The same can be said about higher-order wave mixing of frequencies (so-called three-wave, four-wave, mixing) etc.

We can estimate the coherence length by evaluating the refractive index $n\equiv n(\omega)$ at the relevant frequencies. For instance, in the case of second-harmonic generation, we have $\Delta k=k(2\omega)-2k(\omega)=2\omega[n(2\omega)-n(\omega)]/c$, where we have introduced $k(\omega)=\omega n(\omega)/c$. Hence, perfect phase matching occurs when $n(2\omega)=n(\omega)$, since then

$\Delta k = 0$. In practice we expect $n(2\omega) - n(\omega)$ to be of order 0.01. This gives coherence length of about 50 μm at wavelength $\lambda = 1$ μm.

We can also exploit the non-linearity to perform so-called parametric amplification. This is a process in which power is transferred from a 'pump' wave of frequency ω_3 to waves of frequencies ω_1 and ω_2 (the case $\omega_1 = \omega_2$ is an exact inverse of the second-harmonic generation process outlined above).

9.4. Bistability

One of the most interesting properties of non-linear systems is that they exhibit hysteresis, i.e. that the response of the system to an input light beam depends on the past excitations in the system. This is precisely how a computer memory (or any electronic switch) works. Advances in non-linear optics therefore promise the opportunity of making optical switches that might be used to build optical computers. To demonstrate the essence of the bistability effect in non-linear dielectric media, let us consider a cavity of length l filled with a medium whose refractive index n is a function of the intensity I of the applied optical beam. We shall avoid tedious algebraic manipulations by recording the field dependence of n in a simplified form familiar from optics texts (e.g., the theory of the Fabry–Perot resonator):

$$n = n_0 + n_2 I, \tag{9.40}$$

where n_0 represents the linear response contribution, and n_2 is a constant that is a measure of the field-dependent (non-linear) contribution. The resonance frequency of the cavity is

$$\omega_0 = 2\pi f = \frac{\pi c}{nl}. \tag{9.41}$$

Since n depends on I, ω_0 is also a function of I. We can substitute into eqn (9.41) for n from eqn (9.40), to find

$$\omega_0 = \frac{\pi c}{ln} = \frac{\pi c}{l(n_0 + n_2 I)} = \frac{\pi c}{n_0 l(1 + In_2/n_0)}. \tag{9.42}$$

We know that $n_2 \ll n_0$, so that we can write ω_0 as

$$\omega_0 \simeq \omega_0'\left(1 - \frac{|a|^2}{|a_0|^2}\right), \tag{9.43}$$

where

$$\omega_0' = \frac{\pi c}{n_0 l} \tag{9.44}$$

and $|a|^2/|a_0|^2$ is the normalized intensity such that the amplitude of the wave in the resonator is

$$a = a_0 \exp(\mathrm{i}\omega_0 t - t/\tau_0), \tag{9.45}$$

where τ_0 is a measure of internal losses in the resonator. The amplitude a of the field in the resonator is excited by an incident wave of amplitude

$$s = s_0 \exp(\mathrm{i}\omega t). \tag{9.46}$$

It is customary to introduce τ_e as a measure of an additional decay of amplitude a due to escaping power of the external beam. Then the energy conservation condition for our system is

$$|s|^2 = 2\frac{|a|^2}{\tau_e}. \tag{9.47}$$

The factor two accounts for the fact that waves in both directions contribute to the energy building up in the cavity. The rate of change of a is obtained from eqns (9.45)–(9.47) as

$$\frac{\mathrm{d}a}{\mathrm{d}t} = \mathrm{i}\omega_0 a - (\tau_0^{-1} + \tau_e^{-1})a + \left(\frac{2}{\tau_e}\right)^{1/2} s. \tag{9.48}$$

The response of the system at frequency ω (i.e. when a in eqn (9.48) varies as $\exp(\mathrm{i}\omega t)$) is

$$\mathrm{i}\omega a = \left(\mathrm{i}\omega_0 - \frac{1}{\tau}\right)a + \left(\frac{2}{\tau_e}\right)^{1/2} s, \tag{9.49}$$

where we have introduced

$$\frac{1}{\tau} = \frac{1}{\tau_0} + \frac{1}{\tau_e}. \tag{9.50}$$

Let us substitute for ω_0 from eqn (9.43) into eqn (9.49) and rearrange:

$$a = \frac{(2/\tau_e)^{1/2} s}{\mathrm{i}[\omega - \omega_0'(1 - |a|^2/|a_0|^2)] + 1/\tau}. \tag{9.51}$$

If we multiply this equation by its complex conjugate and divide by $|a_0|^2$, we obtain ($A = |a|^2/|a_0|^2$),

$$A = \frac{2}{\tau_e}\frac{|s_0|^2}{|a_0|^2}\{[\omega - \omega_0'(1 - A)]^2 + \tau^{-2}\}^{-1}. \tag{9.52}$$

It is customary to plot the normalized amplitude A versus B where B is defined as

$$B = \left(2\frac{\tau^2}{\tau_e}\right)\frac{|s_0|^2}{|a_0|^2}, \tag{9.53}$$

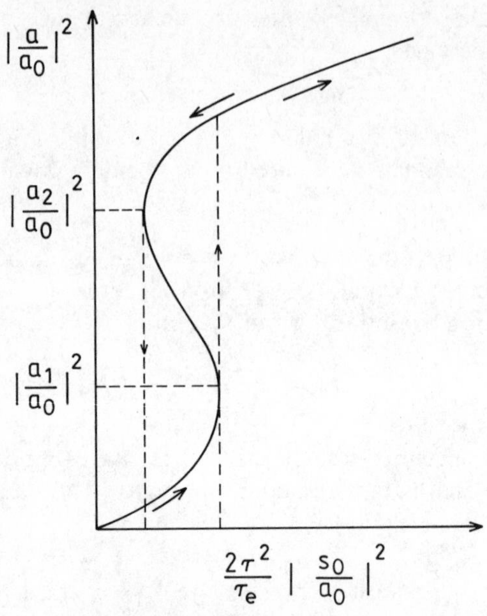

Fig. 9.2. Hysteresis as described in the text, which is characteristic of a resonator filled with a non-linear medium.

for different values of ω. Such a plot is shown in Fig. 9.2. Note that B can be expressed as

$$B = A\{1 + [\omega - \omega_0'(1 - A)]^2\tau^2\}. \tag{9.54}$$

The non-linearity is accounted for by the quadratic function in square brackets. As a result, the curve exhibits a hysteresis in the frequency range

$$\omega < \omega_0'; \qquad (\omega_0' - \omega)\tau > 3^{1/2}. \tag{9.55}$$

The first condition in expression (9.55) implies that the driving frequency ω must be smaller than the cavity resonance frequency ω_0'. The second condition requires that the frequency difference be large enough to overcome the resonator band width.

As we increase the input intensity $|s_0|^2$, the intensity of the field in the resonator first rises to a critical value $|a_1|^2/|a_0|^2$, and then jumps vertically to the upper part of the curve and continues to rise. On the way down the intensity decreases to $|a_2|^2/|a_0|^2$, it then jumps vertically down and continues to decrease to zero. The system only stores information if there is power applied to the resonator. This means that power is continuously dissipated both in the'hold' and 'switch' modes of operation. Only one bit is stored per resonator. The characteristic response time (switching speed) is determined by the response time of the medium (the non-linear dielectric material filling the cavity) or the resonator (whichever is longer). The larger the non-linearity n_2, the smaller is the power needed to operate the device. The power required to switch the device from one state to another is also reduced by reducing the absorption coefficient of the dielectric and the cross-sectional area of the device.

9.5. Self-focusing and soliton propagation

When a beam of finite transverse dimensions travels through a non-linear medium, the refractive index within the volume of space occupied by the beam is different from the refractive index of the medium outside the beam. This is because the non-linear component of polarization depends on the intensity (I) of the applied beam. Suppose that I has a Gaussian profile. This means that at the centre of the applied beam the intensity is larger than the intensity of the beam at points lying further away from the centre. As the beam propagates, it gets progressively more confined to the central region (increasing I implies increasing n, and the centre of the beam travels through the non-linear medium with decreasing velocity ($v = c/n$)). The non-linearity causes self-focusing of the propagating beam (Fig. 9.3).

A beam of finite dimensions also undergoes diffraction. The smaller the cross section of the beam, the stronger the diffraction effect on the shape of the beam profile. Since the diffraction affects mainly the outer part of the beam, its influence is to defocus the beam. It follows that

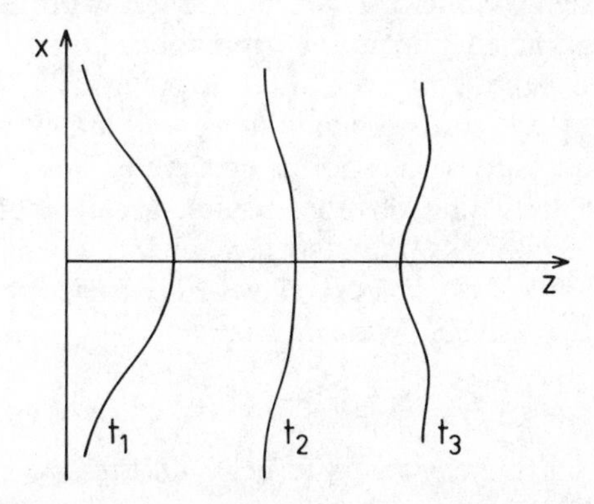

Fig. 9.3. The profile of a beam of light propagating in a non-linear medium along the z direction, at subsequent times t_1, t_2, and t_3. The central portion, where the incident beam intensity is high, propagates more slowly and the beam profile becomes distorted.

under carefully prepared conditions it is possible to balance the effects of self-focusing and diffraction so that the beam profile remains undistorted over a significant distance. Such 'solitary' waves or solitons have been observed in high-quality optic fibres.

9.6. Enhancement of non-linear response in quantum well structures by band filling

The polarization properties of bulk semiconductor crystals are well known. Although the magnitude of the non-linear contribution to the refractive index varies from crystal to crystal, it is generally too small to be of practical use in microelectronics. When the non-linear term is weak, the input power—and consequently heat dissipation in the material—required to achieve a useful bistable element or a mixed wave of significant intensity is far too large. The conventional optical devices exploiting non-linear optical properties of natural materials are large, slow, and consume a lot of power. However, the advent of semiconductor microstructures offers a possibility of creating artificial materials with strongly field-dependent (non-linear) response.

In Section 9.2 we derived the classical expressions for the first- and second-order susceptibility in terms of the characteristic resonance frequency ω_0 of the system. We explained that if we had approached the problem in a full quantum-mechanical treatment, we would have recovered a similar expression, in which ω_0 is replaced by the energy difference (divided by the Planck constant, $\hbar$) between the empty and occupied states, e.g. in the case of the linear (first-order) term $\chi^{(1)}$ by the band gap of the semiconductor material. Hence, with this interpretation in mind, the classical and quantum expressions for $\chi^{(1)}$ are to a good approximation the same. However, the energy levels in a solid must also be characterized by the corresponding density of states $\rho(E)$, and by the probability that the state in question is occupied. This procedure is quite familiar from the discussion of other topics, for example in Chapter 4. We say that the probability that the lower-lying level E_i is occupied is $f(E_i)$. The probability that the upper level E_j participating in the formula for $\chi^{(1)}$ is empty is $1-f(E_j)$. We can write

$$\chi_q^{(1)} \simeq \chi^{(1)} f(E_i)[1-f(E_j)]w_{ij}\rho(E_j), \tag{9.56}$$

where w_{ij} is the transition probability between the two states i and j (w_{ij} also depends explicitly on the wave functions associated with these states; however, we have always assumed that w_{ij} is either 1 or 0 and ignored it.) Hence, although our expression for the first-order susceptibility is independent of the applied field, we can still make $\chi_q^{(1)}$ in eqn (9.56) field dependent by exploiting the properties of $\chi_q^{(1)}$ near the resonance frequency and the role of the distribution function $f(E)$.

The essence of this so-called band filling effect is as follows. If we shine on the material a beam of light whose energy is somewhat larger than the forbidden gap, we excite electrons into the conduction band. If the beam is sufficiently intense, it will maintain a significant population of the conduction band states. We then apply another beam at a frequency that we want to be our operational frequency. To assess the response of our systems at the operational frequency, we invoke eqn (9.56).

The holding beam alters the population of the states near the band edges, in that there are fewer occupied levels at the top of the valence band, and some of the conduction band levels are filled. Consequently, certain excitations across the gap cannot take place, either because there are no valence electrons at the top of the valence band available to make the jump, or because the conduction band states into which the jumping electron is expected to go are occupied. This means that the strength of the effect of the excitation process that, according to our model of Section 9.2, determines the response function is controlled by the intensity of the applied field.

For this mechanism to function efficiently, we must have a system with large absorption coefficient, i.e. a system in which the difference between the absorption (or emission) with and without the holding beam is large. Such a system is best achieved if we use a quantum well structure of GaAs and $Ga_{0.7}Al_{0.3}As$.

In bulk GaAs, the optical spectrum near the band gap is dominated by a sharp excitonic peak. An analogous excitonic peak is observed in GaAs multi-quantum well structures (MQW, Chapter 7). We know that in bulk GaAs the exciton binding energy is so small that at room temperature the exciton is thermally ionized and the advantage of a strong absorption peak is lost. However, in MQW structures the exciton binding energy is increased by the effect of confinement and excitons remain stable even at room temperature. Furthermore, the electrons and holes freed upon exciton dissociation are kept in the active part of the device, i.e. in the GaAs layers, because of the confining effect of the band offset (barrier). This means that the probability of electrons finding holes remains large. The room-temperature absorption spectrum of a typical MQW structure is shown in Fig. 9.4. We can see that the main features of this spectrum are those predicted from our model developed in Chapter 7. We are, as before, interested in the sharp feature at the threshold.

Let us apply an intense (holding) laser beam of energy about 30 meV above the principal gap of the (intrinsic) MQW structure, and measure the absorption of the sample using another source with variable photon energy. If the duration of the holding beam pulse is longer than the

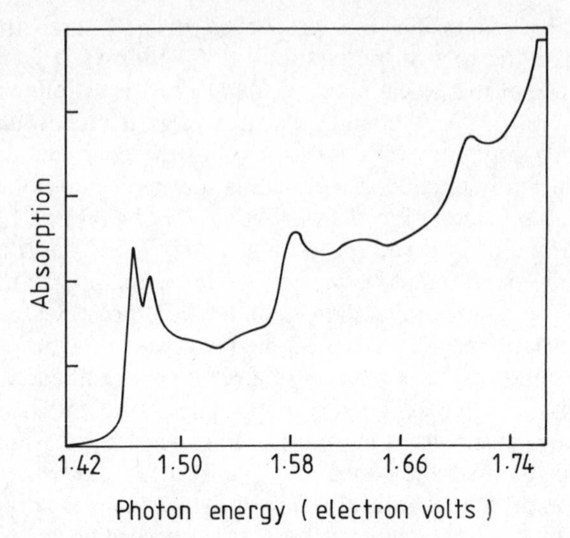

Fig. 9.4. The observed absorption signal of a MQW GaAs–$Ga_{0.7}Al_{0.3}As$ system as a function of photon energy.

average time it takes to excite excitons, we end up with highly energetic electron (and hole) plasma filling the first electron (and hole) minibands. This means that the quantum states that normally supply valence electrons and empty slots in the conduction band for the excitonic absorption to occur are no longer available. Hence the strong excitonic absorption peak (Fig. 9.4) observed when the holding beam is absent is bleached, and the absorption spectrum of the sample looks flat at threshold energies.

The experimental result is shown in Fig. 9.5. The solid line shows absorption with the holding beam on; the broken line is the spectrum without the holding beam (i.e. approximately that of Fig. 9.4 except for a change of scale). Expressed in terms of non-linear susceptibilities, the contrast between the solid and interrupted lines achieved in Fig. 9.5 is equivalent to an increase of the third-order susceptibility $\chi^{(3)}$ of GaAs by six orders of magnitude. The speed of response of the MQW system is given by the speed of electron–hole recombination across the gap (the exciton dissociation), which is of order 1–0.1 ns. Since the magnitude of the contrast in Fig. 9.5 depends on the degree of exciton confinement, it is clear that further progress might be achieved by seeking microstructures in which the confinement is stronger. For example, confinement is greatly enhanced in quantum wires and dots, where the exciton wave function is confined in respectively two and three dimensions, and the radius of the exciton orbit is drastically reduced.

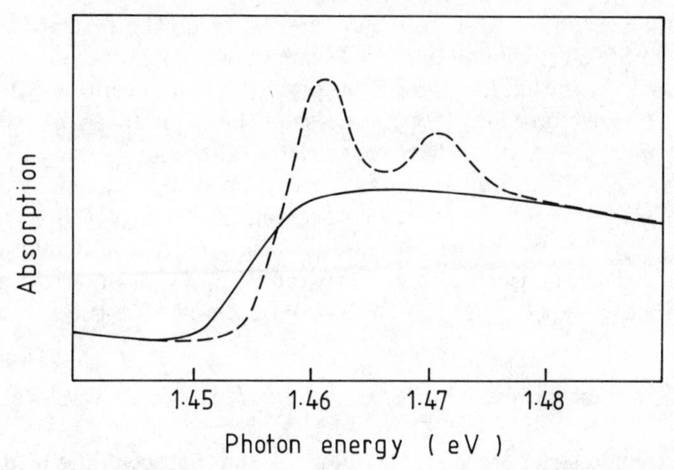

Fig. 9.5. The absorption signal of a MQW system near threshold. The solid line corresponds to the state of the system with the holding beam of energy greater than the band gap energy on, and the broken line is the signal when the holding beam is switched off.

The change of dimensionality also ensures that excitons are more stable and can, therefore, withstand high electric fields. Since an exciton consists of a negatively charged electron and a positively charged hole bound together by the attractive Coulomb force, an external electric field tends to tear the pair apart. This is illustrated in Fig. 9.6. In Fig. 9.6a we have a sketch of the electron and hole wave functions confined in

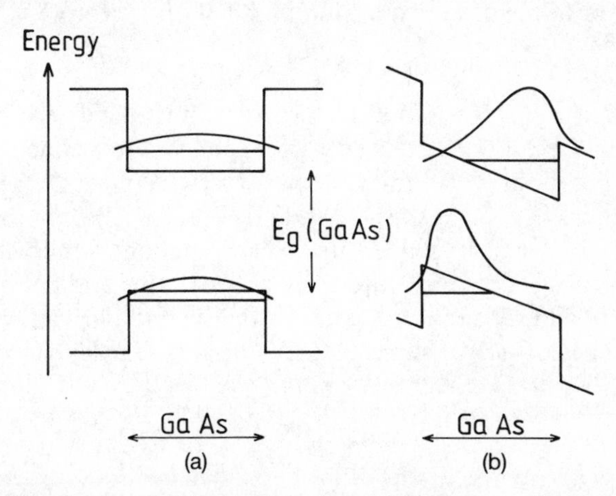

Fig. 9.6. (a) The forms of the wave functions of electron and hole states confined in a GaAs quantum well. (b) The effect of an external electric field, applied perpendicular to the interfaces, upon the shape of these wave functions.

the GaAs layer in the absence of an external field. The effect of the field is to make the electron and hole wave functions peak at the opposite ends of the well, in the region where the potential is more attractive for the particle in question (Fig. 9.6b). The same happens to an electron and hole forming an exciton. This reduces the attraction between the two particles, i.e. the exciton binding energy becomes smaller. However, because of the confining barrier, excitons survive very strong fields (10^5 V cm^{-1}) that are in fact comparable with the exciton binding energy. In a bulk material, such an exciton would rapidly disintegrate and the electron and hole would diffuse away from each other in the strong field.

9.7. Non-linear optical properties of semiconductor superlattices

The electronic structure of semiconductor superlattices differs from that of MQW structure considered in Section 9.6 in that the confined levels are broadened into minibands. The degree of this broadening and the details of the miniband structure (curvature, width) depend on the barrier height (the band offset), on the width of the wells and barriers, and on the effective mass at the band edge of the semiconductor forming the wells. We shall exploit this tunability of the miniband structure in superlattices to enhance the non-linear response of the material.

To understand the essence of the enhancement effect, let us begin by considering an electron at the bottom of the conduction band of a bulk semiconductor like GaAs. Let us measure the electron energy from the bottom of this band at $k = 0$. We can write this energy as a function of wave vector in terms of an expansion at $k = 0$ of the form

$$E = Ak^2 + Bk^3 + Ck^4 + \cdots . \tag{9.57}$$

In a symmetric band, the odd powers of k must have zero coefficients, so that $B = 0$. Recall that we used such expansion to define the effective mass at the band edge of a microscopic crystal. We set $A = \hbar^2/2m^*$ and assumed that close enough to the band edge (at $k = 0$) only the first term is important. That way we ended up with a parabolic band whose energy was $E = \hbar^2k^2/2m^*$. The only difference between this and energy of a free electron is the difference in mass, i.e. in the second derivative of E_k with respect to k. Let us now assume that C is finite and calculate the electron momentum. We find

$$m^*v = \frac{m^*}{\hbar}\frac{\mathrm{d}E}{\mathrm{d}k} = \hbar k + 4m^*C\frac{k^3}{\hbar}. \tag{9.58}$$

It is clear that the non-parabolic contribution to E is responsible for the electron velocity being a cubic function of k. If we apply an electromagnetic field of energy $\hbar\omega$ less than the band gap we expect that the electron momentum will follow the field, i.e. that the electron will oscillate with the frequency of the applied field ω. However, because of the non-linear relationship between v and k, the induced current will also contain mixed frequency components. For instance, if we use two beams of frequencies ω_1 and ω_2, the current may contain components $2\omega_1 - \omega_2$ or $3\omega_1$, etc. In the language of Section 9.2, the crystal behaves as expected of a system whose third-order susceptibility is finite. The band non-parabolicity is therefore equivalent to a non-linearity in the system's response to an external beam.

We can attempt to enhance the band non-parabolicity by 'engineering' artificial structures (e.g. superlattices) for which the non-parabolicity is large. In bulk semiconductors, the non-parabolic term of eqn (9.57) is of order $\delta E^2/E_g$, where $\delta E = (\hbar k)^2/2m^*$. This result can readily be obtained in quantum mechanics by second-order perturbation theory. It is possible to understand this estimate in qualitative terms. Since the constant C in eqn (9.58) is a measure of the deviation of E from the simple parabolic behaviour, it means that as we move (by δE) from $k = 0$ there are additional interactions between the quantum levels in our band structure due to the crystal potential that we have not taken into account in constructing a parabolic band near $k = 0$. The bigger the separation between bands, the weaker is the interaction between them. Hence, C must be inversely proportional to the magnitude of the forbidden gap E_g. This also tells us that semiconductors whose band gap is small must exhibit larger band non-parabolicity. Clearly, for the microstructure to be useful, the non-parabolicity of a superlattice miniband must be substantially larger than the above-mentioned bulk band non-parabolicity.

Having identified the means of generating systems with enhanced non-linear response, we can estimate the relationship between the magnitude of this enhancement and the parameters determining the structure of a semiconductor superlattice. Let us consider an electron at the bottom of the lowest conduction miniband of a GaAs–$Ga_{0.7}Al_{0.3}As$ superlattice of period d. If the separation between wells is not too small, the form of the miniband dispersion (the energy as a function of wave vector) can be obtained in the tight-binding approximation of Chapter 3. In the direction perpendicular to the interface, we can write the electron energy as eqn (3.15).

$$E_k = \Delta[1 - \cos(kd)]. \tag{9.59}$$

Here we have chosen Δ to measure the half-width of the miniband. In the direction parallel to the interface, we assume that the band structure is bulk-like and that it can be represented by the bulk effective mass of GaAs. We shall also assume that the non-parabolicity of the bulk band is small compared to the non-parabolicity we are going to achieve in our superlattice system, and we shall neglect its contribution to the non-parabolicity of the superlattice miniband.

The total energy of our electron in the conduction band is therefore

$$E_T = E_k + \frac{\hbar^2 k_{\parallel}^2}{2m^*}, \tag{9.60}$$

where m^* is the bulk effective mass of GaAs and $k_{\parallel}$ lies in the plane of the interface. Any change in the electronic structure, and consequently in the susceptibility, caused by the confinement effect of eqn (9.59) occurs in the direction of the superlattice axis. In the interface plane, the electronic structure is expected to be parabolic and we shall ignore it.

The effective mass of the superlattice in the direction perpendicular to the interface is m_s^*:

$$(m_s^*)^{-1} = \frac{\mathrm{d}^2 E_k / \mathrm{d}^2 k}{\hbar^2} = \Delta\, \mathrm{d}^2 \frac{\cos(kd)}{\hbar^2}. \tag{9.61}$$

The band non-parabolicity is now built into the expression for the effective mass, since m_s^* is a function of k, Δ, and d. The simplest expression for susceptibility as a function of m_s^* is, from Section 9.2,

$$\chi_s = \frac{Ne^2}{\epsilon_0 m_s^* (\omega_0^2 - \omega^2 + i\omega\sigma)}. \tag{9.62}$$

Substituting for m_s^* from eqn (9.61), we obtain for the perpendicular contribution to the susceptibility of our superlattice system

$$\chi_s = \frac{Ne^2}{\epsilon_0 \hbar^2} \frac{\Delta d^2 \cos(kd)}{\omega_0^2 - \omega^2 + i\omega\sigma}. \tag{9.63}$$

Let us write eqn (9.63) as

$$\chi_s = \chi^{(1)} m^* \, \Delta d^2 \cos(kd) \hbar^{-2} \tag{9.64}$$

Hence, the k-dependent (non-linear) contribution to the superlattice susceptibility can be written as

$$|\delta \chi_s| = \chi^{(1)} m^* d^2 \, \delta E \, \hbar^{-2}. \tag{9.65}$$

We have set the electron energy above the bottom of the band, $[1 - \cos(kd)]\Delta$, to δE. This energy measures the degree to which our electron in the band is excited above the bottom of the band by following the applied beam.

In order to estimate the increase in non-parabolicity achieved in our superlattice compared to the non-parabolicity of bulk material, we must find the non-linear component of χ for the bulk material. Using the result for the electron energy introduced in connection with eqn (9.57), namely,

$$E \simeq \delta E + \frac{\delta E^2}{E_g}; \qquad \delta E = \frac{\hbar^2 k^2}{2m^*}, \tag{9.66}$$

we can proceed as in eqn (9.61) by defining a new bulk effective mass $m^{*\prime}$,

$$\frac{1}{m^{*\prime}} = \frac{1}{\hbar^2}\frac{\mathrm{d}^2 E}{\mathrm{d}k^2}. \tag{9.67}$$

Differentiating, we obtain from eqn (9.66)

$$\frac{1}{m^{*\prime}} = \frac{1}{m^*} + \frac{6\delta E}{m^* E_g} \tag{9.68}$$

The non-linear contribution to the susceptibility of the bulk material is then

$$\delta\chi_b = \frac{6\delta E}{m^* E_g}\frac{Ne^2}{\epsilon_0}\frac{1}{\omega_0^2 - \omega^2 + \mathrm{i}\omega\sigma} = \chi^{(1)}\frac{6\delta E}{E_g}. \tag{9.69}$$

The enhancement achieved by employing the superlattice is given by the ratio of the superlattice non-linear contribution to χ and that of the bulk crystal:

$$\left|\frac{\delta\chi_s}{\delta\chi_b}\right| = \frac{m^* d^2 E_g}{6\hbar^2} \simeq 10^2, \tag{9.70}$$

where we have taken $E_g = 1\,\mathrm{eV}$, $m^* = 0.1\,\mathrm{m}$, and $d = 200\,\text{Å}$.

An additional increase in susceptibility is obtained if the frequency ω of the applied beam is chosen so that $\hbar\omega$ coincides with the separation between the lowest and the adjacent higher-lying (empty) miniband. This corresponds to a resonant enhancement, when electrons are transferred by $\hbar\omega \simeq \hbar\omega_0$. This means that we can estimate the magnitude of this effect simply from σ. The term σ, which we introduced into our expression in eqns (9.1) and (9.26), represents the line width, i.e. $\sigma \simeq \Delta\omega$. Hence, there is an additional enhancement of order

$$\frac{\omega^2}{\Delta\omega^2} \simeq 10^3. \tag{9.71}$$

Here we take $\hbar\omega \simeq 50\,\mathrm{meV}$ for the separation of the minibands, and $\Delta\omega = 1\,\mathrm{meV}$. This increase is again quite substantial.

Finally, there is a correction to our estimate of enhancement due to band filling, since in order to benefit from the increase in eqn (9.71) we must pump electrons from the lowest to the upper miniband. As we pointed out in Section 9.6, the susceptibility is proportional to the number of states available for transitions. The number of electrons δN transferred to the upper miniband is, from the energy conservation condition,

$$\delta N = \frac{\alpha I \tau}{\hbar\omega}, \tag{9.72}$$

where α is the absorption coefficient and I is the intensity of the pumping beam; τ is the time required for the electrons to relax back into the lowest miniband. (We have ignored the details of electron distribution as a function of temperature). It takes one or two collisions of the excited electron with the lattice (phonon emissions) to lose the energy gained in the applied field. This process is very fast ($\tau \simeq 0.1$ ps). Hence, the correction in eqn (9.72) is only important at large pumping intensities.

We have seen that, at least in ideal circumstances, the non-linear response of a GaAs–$Ga_{0.7}Al_{0.3}As$-type superlattice can be 5–6 orders of magnitude larger than in bulk GaAs. Our theory also predicts that an even greater improvement over bulk GaAs should be realized in superlattices consisting of semiconductors with smaller gaps (e.g. InSb, where the bulk non-parabolicity is significantly larger than in bulk GaAs). The response time of the system is determined by the time it takes an electron to relax back into the lowest state, which is less than 1 ps. This is several orders of magnitude faster than the band filling mechanism discussed earlier, for which the response time depends on exciton recombination and consequently involves the slower transitions across the principal gap. However, since the separation between minibands is small, for the resonant enhancement to occur the operational frequency must lie in the far-infrared band, which is not a favoured range for optical communication devices.

Problems

9.1. Make a rough comparison of susceptibility χ per electron for a free hydrogen atom and for a silicon crystal, in the limit $\omega \to 0$.

9.2. Make a qualitative comparison of the magnitude of χ of a silicon crystal with that of a simple metal, in the limit $\omega \to 0$.

9.3. Compare the frequency at which the lattice vibration contribution to the first-order response of a solid is significant with that of the electronic contribution.

9.4. Estimate the coherence length for second-harmonic generation in GaAs with $\omega = \frac{1}{4} E_g/\hbar$.

9.5. Consider GaAs–$Ga_{0.7}Al_{0.3}As$ and InSb–CdTe superlattices of period $d =$ 150 Å. Estimate the enhancement of the non-linear part of susceptibility that can be achieved in these structures relative to the bulk non-linearity in GaAs and InSb, respectively. Choose ω away from resonance.

9.6. Estimate the enhancement of susceptibility in the structures considered in Problem 9.5 when the response is due to holes at the top of the valence band.

9.7. Discuss the conditions under which the expression $(\omega/\Delta\omega)^2$ for resonant enhancement of the non-linear contribution to susceptibility is approximately valid. How would the effect of temperature influence the results in eqns (9.71) and (9.72)?

9.8. Consider the limits of applicability of the approach leading to eqn (9.70).

9.9. Consider a GaAs–$Ga_{0.7}Al_{0.3}As$ superlattice of (large) period d in which the doping is chosen so that the lowest conduction miniband is half filled with electrons. Calculate the susceptibility in the manner of eqns (9.61)–(9.64) and compare it with the result obtained when the active electron lies at the bottom of the conduction miniband.

9.10. Compare the contribution (away from resonance) to the susceptibility of an intrinsic GaAs–GaAlAs superlattice from a valence electron and that from an electron placed at the bottom of the lowest conduction miniband.

10

Fabrication and characterization of semiconductor microstructures

10.1. Methods of crystal growth

Semiconductor crystals used in the 1950s and 1960s to make transistors were macroscopic objects that could be handled manually and examined with the naked eye or under some primitive microscope. These crystals were prepared by pulling an ingot out of melted raw material contained in a furnace. The details of the atomic arrangement at the interface between the semiconductor and a metal or insulator did not matter very much, since the volume-to-surface ratio of the semiconductor could be regarded as infinitely large. The electron transport effects and the general functioning of the devices made in those days were dominated by bulk phenomena.

Much of the effort concerning crystal growth and characterization was initially concerned with improvements in crystal purity and control of doping. In the course of this endeavour, a number of techniques have been developed that have since become standard in semiconductor technology, and which are commonly used to characterize any semiconductor material. However, with the advent of semiconductor microstructures, new techniques of crystal growth and characterization have emerged, and it is the purpose of this chapter to outline these new developments.

In semiconductor microstructures, very thin film of perfect crystalline structure—often consisting of only a few atomic layers—must be deposited with great precision on a suitable substrate. Such layers are grown by methods of liquid-phase epitaxy (LPE), molecular-beam epitaxy (MBE) and metallo-organic chemical vapor deposition (MOCVD). In the LPE method, layers of, say, GaAs are grown on a thick slice of high-resistivity GaAs by cooling at prescribed rates a heated liquid solution containing gallium and arsenic. In the MOCVD technique, the substrate is exposed to a hot stream of gaseous compounds, and in MBE to a thermal beam of atoms and molecules. New hybrid techniques have also been employed,

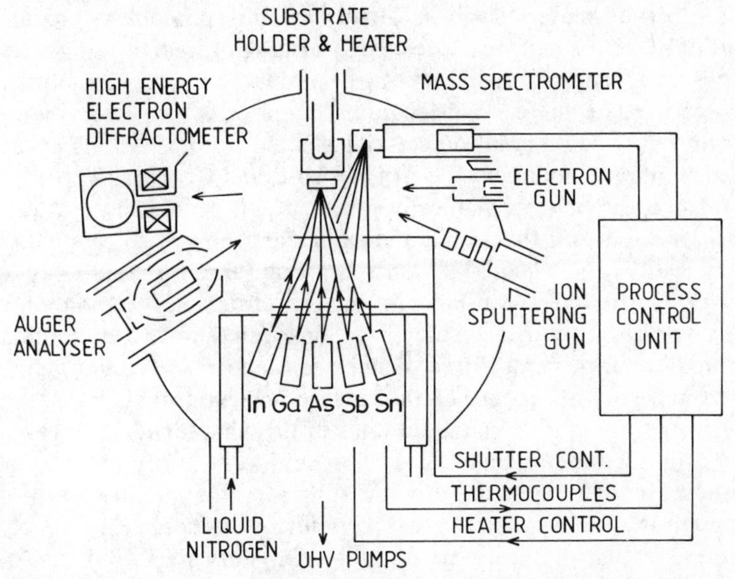

Fig. 10.1. A reactor for molecular-beam epitaxial growth of semiconductor microstructures (e.g., GaSb-InAs superlattice doped with Sn impurities).

in which certain advantages of each of these methods (e.g. operational temperature, speed of deposition, area of deposited material, flexibility in achieving more sophisticated structures such as wires, etc.) are combined while disadvantages are suppressed.

MBE offers probably the best control over deposition and is therefore a natural choice for microstructure studies and development. The core of the reactor consists of a ultrahigh-vacuum chamber (Fig. 10.1). Several evaporation cells, each of them controlled by a separate shutter, supply hot fluxes of molecular beams of the required species (e.g. gallium and arsenic). A separate source (an effusion cell) is used for each host crystal constituent as well as for each dopant. The flux can be initiated or terminated within a tenth of a second. The growth rate is about 5 Å of layer thickness per second. It is possible to reload a substrate into the chamber without breaking the high vacuum that ensures the purity and undisturbed progress of growth. A heavy-ion gun is provided in the chamber. It can be used to bombard the surface and remove unwanted features such structureless conglomerates of atoms on the surface, steps, and other irregularities. After such bombardment, the damaged crystalline layers near the surface are repaired by slowly heating the material (annealing). The substrate is maintained at about 500–700°C, or at as low a temperature as possible so as to minimize diffusion of the deposited atoms.

The substrate material and its quality are indispensable ingredients in the art of making small structures. Ingots of high-resistivity single crystals as large as 150 mm in diameter can be grown that are virtually free of macroscopic defects (e.g. dislocations), and with only very low concentration of residual point defects. This is the result of a long effort that has led to improvements in the original so-called Czochralski procedure. The substrate is first mechanically polished and chemically etched, then quickly inserted into the vacuum chamber and heated to remove oxide. Sometimes it is also cleaned by ion sputtering (argon is commonly used). By a 'clean surface' we usually mean the critical top region of about 5–15 Å, which is so perfect that it retains bulk-like order, contains no unwanted impurities and is free of steps or islands. The Auger analyser is used to monitor the quality of the surface achieved in such a treatment (the Auger technique is discussed later in this chapter). The high-energy electron diffractometer in the chamber monitors the smoothness of the deposited film. The epitaxial growth is initiated by opening the shutters. This process is controlled by a computer. It is connected to a mass spectrometer that monitors the flux rate. Dopants are introduced during growth at desired rates. The system shown in the figure can be used, for example, to grow a GaSb–InAs superlattice doped with tin.

The MOCVD technique of depositing, say, GaAs layers is based on a chemical reaction in which $(CH_3)_3Al$, $(CH_3)_3Ga$, and AsH_3 decompose to form GaAlAs and CH_4. Again, the growth rates and layer quality are computer controlled via flowmeters. Although the MOCVD technique is in general less well suited for making sharp semiconductor–semiconductor interfaces, it is cheaper to operate and yields excellent material.

The MOCVD method and its derivatives therefore offer particularly good promise for industrial applications.

10.2. Lithography

The quality semiconductor wafers grown by one of the high-precision techniques provide raw material for manufacture of integrated circuits and other semiconductor microdevices. To turn a semiconductor wafer into an integrated circuit, it is necessary to transform the uniform crystalline layers into a complicated pattern, with many thousands of transistors connected in a particular manner suitable for further processing and device fabrication.

The first circuits 'integrated' on a single silicon chip were made by chemical etching. The quality of the mask and its fabrication could be checked under a desktop optical microscope. However, new methods are needed in order to achieve the resolution required for application to very small structures. The basic optical lithographic processes employed to make patterns with linewidths larger than about 0.5 μm rely on a light-sensitive layer deposited on the material and later exposed to radiation of suitable wavelength. If, in the subsequent development, the areas exposed to this radiation are removed, the resist is called positive. If the exposed areas are hardened by the radiation and the unexposed region is treated (removed) by the developer, the resist is said to be negative.

For line widths smaller than about 0.5 μm, we need radiation of shorter wavelengths, and an electron beam or X-rays must be used. An electron beam can be focused to a spot of about 10 Å diameter. Machines for making patterns by electron-beam lithography consist of a source of electrons and of a set of 'lenses' to collect the electrons and focus the beam onto the resist. Deflection coils direct the beam under the control of a computer that has been programmed to produce the desired pattern. Accurate movements of such a beam over a large area are difficult to achieve. Various methods have been designed to enable the machine to 'write' patterns quickly and flexibly. For the high-resolution line widths (below 0.1 μm), a modified version of the scanning electron microscope has been employed. Wafers 5 cm in diameter can be processed with this machine with maximum spot size of about $2\text{–}10 \times 10^{-7}$ cm. In these machines, the control of the electron distribution in the beam, and its maintenance during the writing process are of key importance. In the electron beam writing, highly focused electrons hit the surface of the crystal and make elastic (without energy loss) and inelastic collisions with atoms. In an inelastic collision, the energetic electron is stopped and its kinetic energy is transferred to the valence electrons at the surface. This means that a beam of energetic secondary electrons is generated in this process. The secondary electrons can be reflected back to the surface (with an energy different from that of the primary beam). The effect of the back-scattered electrons is to reduce the overall contrast in the latent image formed by electron exposure in the resist. For example, for a 1 μm thick layer of resist on a silicon surface, exposed to electrons of 20 kV energy, the back-scattered electrons increase the line width to 0.15 μm. Using a thin resist (on a thick substrate) makes the back-scattered electrons spread uniformly and the linewidth is reduced to about 5×10^{-6} cm. The effect of back-scattering can be almost completely eliminated by using very thin substrates.

10.3. Dry etching

Another way to replace conventional (wet) chemical etching is to use so-called dry etching. The material is removed either by ion bombard-

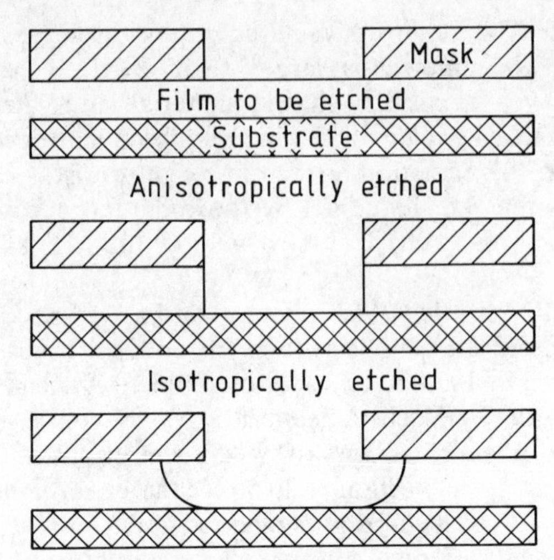

Fig. 10.2. An illustration of the difference between isotropic (wet) etching and anisotropic (dry) etching profiles.

ment or by the formation of gases owing to chemical reaction of the material with a suitable radical. The term dry etching is used in connection with a variety of combinations of these two effects. In general, any etching process in which a solid surface is etched in a gaseous environment is called dry etching. The advantage of the dry

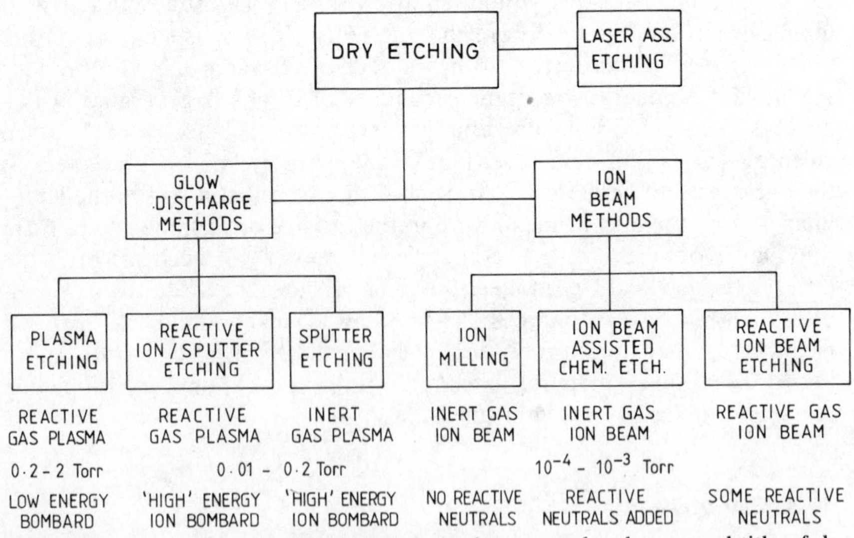

Fig. 10.3. A summary of various techniques known under the general title of dry etching (1 Torr = 1 bar.)

etching process over the wet one is that it leads to an 'anisotropic' edge profile. The meaning of this description is illustrated in Fig. 10.2, where the 'isotropic' and 'anisotropic' profiles in the edged film characteristic of the wet and dry techniques, respectively, are illustrated. The anisotropic profile is clearly preferable.

The broad differences between the various methods linked to the general label 'dry' etching, and headings under which they are referred to in the literature, are summarized in Fig. 10.3. In the glow discharge method, a reactive gas plasma is created in the vacuum chamber where the substrate is located. In techniques using an ion beam, the plasma is generated in a separate chamber. The ions are then directed towards the substrate. In sputter etching, and in ion-beam milling, the atoms are removed only by the momentum transfer between the energetic (argon) ions and the crystal surface. In all other methods, the plasma also has a chemical reaction with the film and this assists the etching process. The figure gives the characteristic pressures for the processes in question in units of Torr (1 Torr = 1 bar).

10.4. Direct writing

One of the most advanced techniques for making patterns on semiconductor films is the laser-induced 'direct writing'. The laser radiation 'spot' can be positioned and moved over the surface with great precision. The irradiation of reactant ambient gases generates activated species in well-defined places and local growth or material removal is achieved without masks (hence the label 'direct').

Figure 10.4 is a sketch showing laser-induced chemical vapour deposition on a substrate. Thus efficiency of the technique depends on several interactions that are a sensitive function of the microscopic physical properties of the gas molecules and of the substrate material. When the laser light is absorbed by the gas, the photons absorbed by the molecules excite the vibrational and rotational molecular motion, which dissociates the molecules. This photolysis of the molecular gas can be well controlled by choosing the frequency and intensity of the laser beam as well as the spot size d. The products of the photolysis are active atomic species that then react with and are deposited on the surface of the solid.

The opposite process can also be used to remove adsorbed molecules from the surface. Since the absorption spectra of the substrate and the molecules forming the gas are in general very different, the frequency of the applied laser beam may be chosen in this approach to excite the substrate selectively. The energy absorbed by the substrate material is transferred locally to the molecules, i.e. only in the locations where this

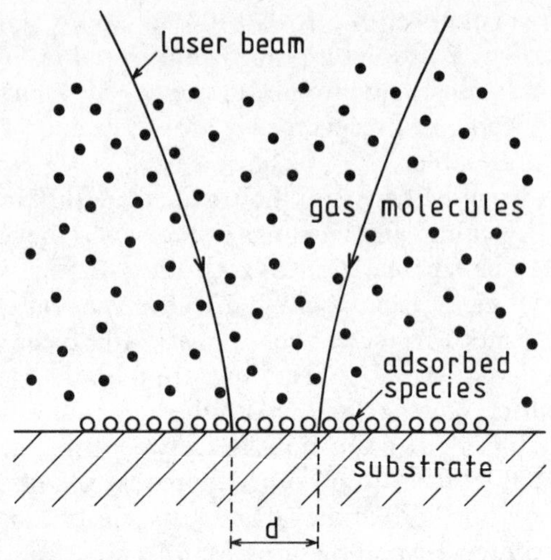

Fig. 10.4. A diagram illustrating the principle of direct writing. A laser beam of diameter *d* heats locally a substrate with adsorbed molecules.

process is desired. The heating leads to thermal dissociation of the adsorbed molecules in the heated areas. In order to achieve high resolution (narrow line widths), the length of the pulse as well as the size of the spot must be well controlled and made as small as possible. This technique is particularly desirable for speedy and accurate fabrication of circuit interconnections.

10.5. Langmuir–Blodgett–Roberts films

The need to manufacture very small structures in a strictly reproducible and economical way requires methods of growth that guarantee a monolayer accuracy in film deposition, and that offer a 'digital' control mechanism to specify the number and chemical character of the atomic monolayer in question. This degree of control has been achieved in monolayer assemblies of organic molecular matter such as stearic acid and its derivatives. Many years ago, such monomolecular layers about 250 Å thick were prepared by I. Langmuir and K. B. Blodgett on the surface of water. More recently, G. G. Roberts has deposited such films of organic insulators on semiconductors. A single molecular monolayer is first prepared on water and then transferred from the water surface onto a substrate by dipping it into the water. On each withdrawal of the substrate, a single monolayer of the molecular insulator is deposited.

The thickness of the monolayer depends on the organic material used and is known with great precision and in advance of the application. These molecules have excellent insulating properties and have been used in device manufacture.

10.6. Digital MBE growth

In conventional epitaxial growth techniques outlined in Section 10.1, the layer thickness increases linearly with deposition time. The growth rate is a complicated function of the temperature of the source and substrate, pressure and flow rates of source gases, and other factors. Under such circumstances, it is difficult to achieve reproducible structures. The running of the machine is also unsuitable for mass production of microdevices, since it requires analogue control programs. However, if the thermodynamic conditions of growth resemble those characteristic of the formation of Langmuir–Blodgett–Roberts films, it is also possible to grow semiconductors in one-monolayer steps, i.e. such that the probability of double adsorption of the deposited species is negligible. To ensure that the growth resembles the desired step-like sequence, the operational pressure P and the adsorption coefficient K of the deposited material on the substrate in question must satisfy the inequality $KP \gg 1$. The MBE growth sequence for, say, ZnS might then be arranged in steps as indicated schematically in Fig. 10.5. The open and filled circles indicate the two species, zinc and sulphur. The build-up of a semiconductor microstructure could then be expressed in a simple digital code.

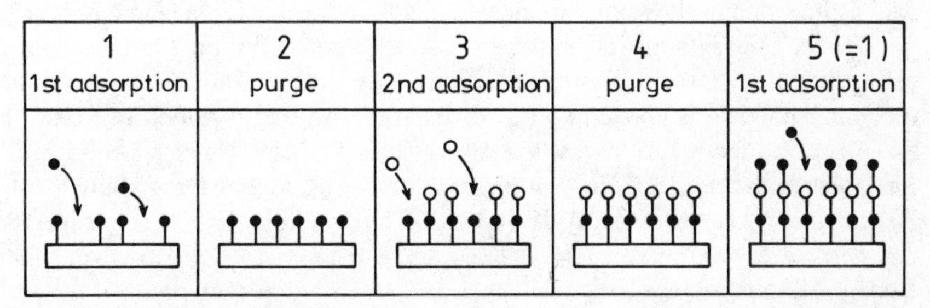

Fig. 10.5. A sequence of monolayer growth of a III–V or a II–VI semiconductor. For example, taking ZnS, we have: Stage 1, zinc vapor is supplied and adsorbed on the (111) sulphur interface plane of the substrate ZnS material; State 2, zinc coverage is complete and the remaining vapor is removed; State 3, sulphur vapor is introduced; State 4, excess sulphur is removed; State 5, the beginning of next cycle (as in Stage 1).

10.7. Surface spectroscopy

The geometrical arrangement of atoms on surfaces is an important means of assessing the surface properties, and for establishing the above mentioned condition of 'clean surface' required for successful preparation of high-quality material. This arrangement of atoms is studied in a high-vacuum chamber by low-energy electron diffraction (LEED). Low-energy electrons are chosen for the task because their wavelength λ is comparable to the internuclear separation of atoms on crystal surfaces. From the so-called de Broglie relation, linear momentum $p = mv = \hbar k = 2\pi\hbar/\lambda$. If the electron energy is E, we have $\lambda(\text{Å}) \simeq [150/E(\text{eV})]^{1/2}$, so that electrons with energy 150 eV have wavelengths of 1 Å. Because of their low energy, these electrons do not penetrate deep into the crystal and their Bragg diffraction pattern corresponds to diffraction at the surface layer. The incident angle is often chosen to be very small (about 3° only) so that somewhat higher energy (about 30 kV) can be used. The process is than called reflection high-energy electron diffraction (RHEED). The low angle increases the sensitivity of the method and provides information about small displacements of surface atoms from their bulk positions. Such displacements, or surface reconstruction, although very small compared to the separation between atoms, strongly affects the way atoms can be adsorbed on the surface and represents an important piece of information for establishing the correct growth conditions. The RHEED facility is usually built directly into the MBE reactor as a standard characterization tool.

The presence of impurities on clean surfaces is detected by Auger spectroscopy. The principle of the Auger effect in indicated in Fig. 10.6. Consider a single impurity atom on an otherwise perfect surface. In a free atom, electrons occupy energy levels labelled K, L, M, etc., K being the lowest energy state occupied by two electrons. Suppose now that an energetic electron or photon hits such an atom. The energy of the incoming particle is chosen so that it is large enough to knock out one of the electrons from the lowest shell (e.g. the K shell). Such a situation is unstable, because electrons residing in one of the higher lying shells (e.g. L) can jump into the empty K slot. The energy released in this process $\Delta E = E_L - E_K$ may be carried off by an emitted photon. However, the energy ΔE can also be disposed of in the so-called Auger process. In the quantum-mechanical picture of a many-electron atom, electrons are held together by strong forces to which all electrons contribute, i.e. the energy of the system is shared by all particles. Therefore, the ΔE that must be released in order to remove an electron from the L shell and place it into the empty slot in the K shell can be taken up by one of the higher-lying electrons, say in shell M. This electron is then emitted (released from the atom) with the characteristic kinetic energy $\Delta E -$

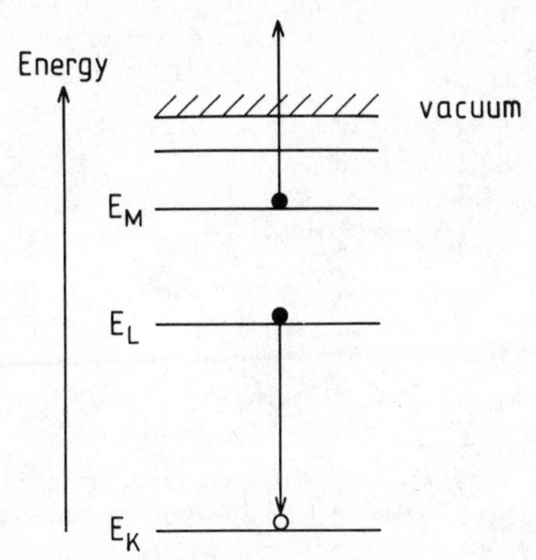

Fig. 10.6. The Auger process, in which an electron is first excited by a highly energetic beam from a free-atom core shell (K). The empty place is then filled by an electron from the upper L shell. The excess energy is given to an electron in shell M, which is released from the atom.

$|E_M|$. Both ΔE and E_M are known very accurately from earlier studies of free-atom spectra. Thus each atom can be associated with a characteristic 'Auger' spectrum. A coverage of only one impurity atom per thousand host atoms can be detected this way. Auger spectroscopy is thus a powerful tool for checking surface purity.

10.8. Quantum Hall effect

One of the commonest means of characterization of semiconductor crystals is measurement of the Hall effect. The Hall effect is concerned with the motion of a free carrier (e.g., an electron at the bottom of the conduction band of a semiconductor material) under the influence of crossed electric ($E_x \parallel x$) and magnetic ($B \parallel z$) fields, as indicated in Fig. 10.7. Under the electric field, the carrier is influenced by a force $m^*(\mathrm{d}v_x/\mathrm{d}t + v_x/\tau)$, where v is the (drift) velocity and τ is the relaxation time (the average time between collisions). The effect of the magnetic field is to exert the Lorenz force $e(\mathbf{E} + \mathbf{v} \times \mathbf{B})$. Thus, for each component of velocity v_x, v_y, and v_z we obtain an equation of motion:

$$\begin{aligned} \dot{v}_x + \frac{v_x}{\tau} &= -\frac{e}{m^*}[E_x + Bv_y] = -\frac{e}{m^*}E_x - \omega_c v_y; \\ \dot{v}_y + \frac{v_y}{\tau} &= -\frac{e}{m^*}[E_y - Bv_x] = -\frac{e}{m^*}E_y + \omega_c v_x; \\ \dot{v}_z + \frac{v_z}{\tau} &= -\frac{e}{m^*}E_z. \end{aligned} \tag{10.1}$$

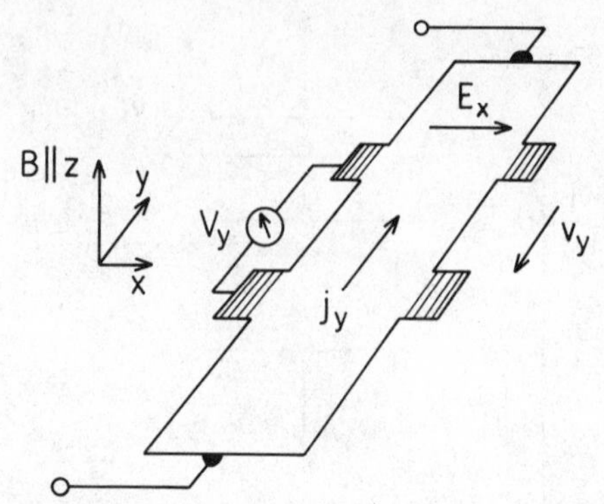

Fig. 10.7. A sample of bar-shaped geometry and an arrangement of electric and magnetic fields for Hall effect measurements.

The effective mass of electrons (or holes) at the band edge in a semiconductor material differs substantially from the free-electron mass and we have therefore introduced the effective mass m^*.

Under the influence of B, the carrier moves along a circle with the plane of orbit perpendicular to the field with a characteristic ('cyclotron') frequency $\omega_c = eB/m^*$. This means that when the electric field in the x direction is switched on, the electrons experience a force deflecting them along the y axis. This continues until the electric field, generated in the y direction owing to the deflected current, and acting against the effect of the Lorenz force, is in perfect balance with it. In the steady state, v_x is constant, the (average) velocity in the y direction is zero, and we can compute from eqn (10.1) the magnitude of E_y:

$$E_y = -\omega_c \tau E_x. \tag{10.2}$$

The Hall resistivity is defined as $R = E_y/j_x$, where $j_x = -euv_x$, and u is the number of carriers per unit area. Substituting for E_y and j_x, we obtain

$$R = -\frac{B}{eu}. \tag{10.3}$$

The effect of the magnetic field can also be exploited in the absence of an applied electric field. Thanks to the Lorenz force, the electron motion in the x,y plane is circular with angular frequency ω_c, so that the electron energy is that of a two-dimensional harmonic oscillator. In the z direction the electron energy is free-electron-like so that we can write the total electron energy E as

$$E = \frac{\hbar^2 k_z^2}{2m^*} + \hbar\omega_c(l + \tfrac{1}{2}), \tag{10.4}$$

where l is 0, 1, 2, . . . ,. The electron energy can be divided into one-dimensional sub-bands or Landau levels, each of which is associated with a particular value of l. We say that the energy is quantized in the x,y plane. The separation between sub-bands is a multiple of $\hbar\omega_c$. It follows that at low temperatures (when $k_BT < \hbar\omega_c$), and when the mean free path is larger than the length of the electron orbit, this quantization of energy becomes observable. If the lth sub-band is the highest populated level, then it is possible to excite electrons into the nearest empty sub-band by supplying electromagnetic radiation of energy $\hbar\omega_c$. This is done in a cyclotron resonance experiment in which one looks, for instance, for frequencies at which absorption occurs for a fixed value of the magnetic field. For magnetic fields of the order of several teslas, the cyclotron frequency falls into the microwave band. At 10 T we obtain in silicon $\hbar\omega_c \simeq 6$ meV, whereas in GaAs $\hbar\omega_c$ is about 17 meV because of the smaller effective mass. Since ω_c is inversely proportional to the effective mass, the empirical value of ω_c can be used to determine directly the magnitude of m^*.

Let us now consider what happens when the width of the sample in the z direction is reduced so that it becomes small or comparable to the carrier de Broglie wavelength. This can be achieved, for example, in inversion layer structures. We know from Chapters 6 and 7 that under such circumstances the carrier energy in the z direction no longer resembles the continuum of eqn (10.4). Instead, it consists of discrete levels. Thus the combined effect of confinement and magnetic field turns the number of states per unit area (u in eqn (10.3)) available to electrons into a step-like function. The quantization of the oscillator states implies that $\Delta x\, \omega_c m^* = h$. Substituting for ω_c, we find that there are eB/h oscillator states per unit area. If the M lowest states are occupied, the Hall resistivity of eqn (10.3) becomes

$$R = -\frac{B}{eu} = -\frac{h}{e^2 M}, \tag{10.5}$$

where $M = 1, 2, 3, \ldots$ is an integer indicating how many Landau levels are filled. By varying the applied field, we can observe the step-like rise of the Hall resistivity, which changes every time the Fermi energy passes through a Landau level. The spectrum is characteristic of the two-dimensional nature of the system.

In three-dimensional (bulk) samples, the Hall resistivity obtained in eqn (10.3) is a continuous function of B and the electron density. Although the effect of the magnetic field leads to quantization of the oscillator states in the x,y plane, the density of states in such a system is a continuous function of energy. Note that the density of states of the one-dimensional continuum represented by $\hbar^2 k_z^2/2m^*$ in eqn (10.4) varies as the square-root of energy, i.e. in the same way as for a three-dimensional continuum discussed in Chapter 1.

The first observation by K. von Klitzing of the existence of the step-like character of Hall resistivity in quasi-two-dimensional systems—which is known generally as the *quantum Hall effect*—was carried out on MOS-type semiconductor structures. However, analogous results can be obtained in other heterojunction systems.

10.9. Determination of heterojunction band line-ups

Heterojunctions are basic building units of semiconductor microstructures. In particular, the key parameter on which much of the physics of low-dimensional structures depends is the heterojunction band offset. An accurate value of the band offset is required to determine the position of confined levels, and consequently the band gap of a microstructure. It also determines the degree of penetration of electron wave functions into the barrier.

The magnitude of the band offset depends on the relative position of the bulk band structures of the constituent materials. However, the actual value detected in an experiment may reflect the degree of contamination by impurities, the interface roughness (irregularities), inhomogeneous distribution of atoms forming an alloy, or fluctuation in doping concentration. The observed value is therefore likely to vary from sample to sample. An additional uncertainty is also introduced by errors peculiar to the experimental technique itself (finite line width, uncertainty in the actual value of the electric field in the sample, etc.).

There are several obvious ways of using spectroscopic information about a semiconductor microstructure to determine the band line-up. One is to invoke the principle of thermionic emission outlined in Chapter 6. In a structure exhibiting a barrier (in this case the barrier is our band offset) of height W, the current density j is proportional to $T^2 \exp(-W/k_B T)$, where T is temperature. We can measure j as a function of T and from the slope of $\ln(j)$ determine the magnitude of W.

Another electron transport effect that can be exploited is tunnelling. If we apply an electric field perpendicular to the interface plane, we can use the WKB expression of Chapter 8 for the tunnelling current at low temperature, which relates it to the barrier height W. The difficulty with such electron transport measurements is that they are not selective to any particular transport process. The observed current is the sum of all contributions to the current occurring between the two electrodes and it is bound to be affected by the details of the space charge potential, the fluctuation of the sample resistivity, impurities, and other factors unrelated to the band line-up. The magnitude of W then depends on the accuracy of the theoretical model that must be used to provide input into the formula used in the calculation of W from the observed value of current. We have seen that the band offset is often a small quantity, typically of the order of several tenths of an electron-volt. The error in the above procedures is likely to be of the order of 0.1 eV.

It is also possible to estimate band offsets from optical data concerning the position of confined levels in, say, quantum well structures. In our discussion of confinement, we used the particle-in-a-box picture to determine the position of quantum states in wells whose depth is equal to the conduction or valence band offset ΔE_c and ΔE_v, respectively. If we obtain the empirical value of the forbidden gap from absorption or luminescence experiments, we can seek the best values of ΔE_c and ΔE_v to fit the data. However, even if we assume that no contamination by impurities is involved, the interfaces are sharp, and there is no space charge effect, we encounter a number of difficulties that diminish the accuracy of such methods of assessment. The spectroscopic data provide us with the magnitude of the band gap, i.e. with the energy *difference*, so that we can only compare the sum of the energies of the confined levels in the wells with the empirical data. If we increase ΔE_c in our fitting procedure, we must decrease ΔE_v, since their sum must be equal to the difference between the bulk band gaps of the constituent semiconductors. By increasing ΔE_c we are increasing the separation of the confined levels from the conduction band edge. However, we decrease ΔE_v, and therefore decrease the separation of the levels near the valence band from the band edge. This means that the band gap energy observed in an optical experiment is insensitive to the change in the ratio of the band offsets ΔE_c and ΔE_v. This is illustrated in Fig. 10.8, which shows two

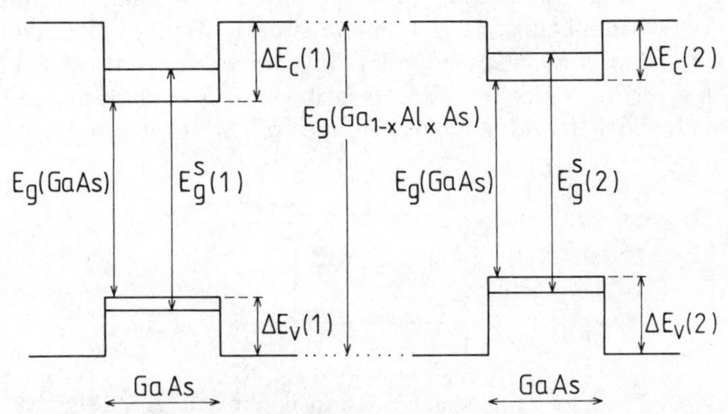

Fig. 10.8. Two band diagrams, showing band edges for a quantum well structure constructed with different band offsets ΔE_c and ΔE_v. E_g^s is the superlattice gap.

band diagrams for a quantum well of GaAs, each constructed with a somewhat different band offset ratio. The observed gaps in the two cases, $E_g^s(1)$ and $E_g^s(2)$, and the band offset, ΔE_c and ΔE_v, are indicated.

Since the sum of band offsets is a constant given by the difference in bulk gaps, it is customary to quote the ratio of the conduction and valence band offset, $\Delta E_c/\Delta E_v$. This helps to express explicitly any change in the way the two bulk band structures are positioned relative to each other. For example, it we use room-temperature bulk gaps as a starting point, we can say that the band offset of a GaAs–AlAs heterojunction is given by the ratio 70/30. (It is important to remember that the band gap of a semiconductor decreases with temperature by about 0.5 meV/K, so that the above ratio depends to some extent on temperature.)

The accuracy of the band offset determination from optical data can be improved if transitions (gaps) separating second, third, etc., minibands are included into the analysis. We have argued earlier that the 'selection rule' for observable transition is $\Delta l = 0$, where l is an integer (a quantum number) that labels confined levels starting from the bottom of the well in question. Hence, under favourable conditions, transitions can be seen between, say, the second conduction and second valence levels. The positions of the higher-lying levels are in general more sensitive to the magnitude of the barrier (band offset) and consequently improve the chances of getting a more accurate assessment.

Further improvements in the effort to determine band offsets can be achieved if the data concerning the forbidden gap are supplemented by observations of transitions from one conduction (or valence) level to another. For example, we can take electrons from the lowest level to the next above it, or to the level lying third above the bottom of the well, as indicated in Fig. 10.9. This provides direct information about the separation of minibands. It is then possible to fit the conduction and valence band offsets independently. The separation between adjacent confined levels in the well is more sensitive to the height of the barrier and also to the well width. However, the well width is often not known

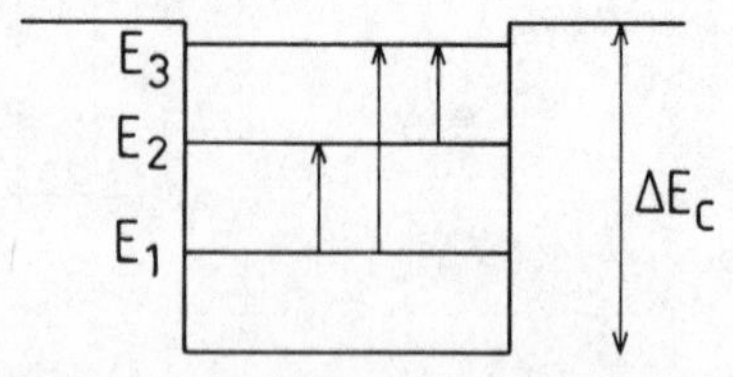

Fig. 10.9. Confined levels 1, 2, and 3 in a quantum well. ΔE_c is the conduction band offset. The arrows indicate inter-sub-band transitions. The level separation is typically of order 10–50 meV.

very accurately. The level depth in the well varies as the square of the well width, so that an error of 5 Å (which is the minimum experimental error) in the well width may represent a significant correction in the fitting procedure, particularly when narrow wells are involved. Furthermore, inter-sub-band transitions such as those indicated in Fig. 10.9 are difficult to measure because they lie in the far-infrared range of energies, where tunable sources of radiation are scarce, and the absorption signal is often weak and obscured by extrinsic effects (such as impurity absorption). Also, in order to see such transitions, the material must be doped so as to populate the miniband from which the absorption transition is to originate. The electric field due to the space charge alters the positions and the width of the levels confined at heterojunctions and the resulting line broadening reduces the accuracy of the level separations determined in such experiments.

Finally, the magnitude of band offsets can be determined by invoking the pressure dependence of the semiconductor bulk band structure and its manifestations in low-dimensional systems. Consider, first, a bulk GaAs crystal. If we apply uniform hydrostatic pressure, and plot the magnitude of the separations (gaps) between the top of the valence band and the conduction band minima that we labelled in Chapter 5 as Γ (the principal, lowest minimum), X and L (secondary minima at the Brillouin zone boundaries along the $\langle 100 \rangle$ and $\langle 111 \rangle$ directions, respectively), we obtain a diagram such as is shown in Fig. 10.10. The rates of change of these energies as a function of applied pressure, the so-called pressure coefficients, have been tabulated for most technologically important semiconductor materials. We can see in Fig. 10.10 that the lowest Γ minumum moves rapidly up and the (direct) gap of GaAs increases with pressure by about 11 meV/kbar. The secondary X minimum moves down in energy, at a rate of about 1.24 meV/kbar. This means that the (indirect) gap of GaAs between the X conduction band minimum and the top of the valence band (which occurs at the centre of the Brillouin zone at Γ) decreases until it crosses below the direct gap (Γ) value at around 40 kbar. It is interesting to notice that the band structure of GaAs under high pressure ($\simeq$60 kbar) in fact strongly resembles the band structure of another semiconductor, GaP (at atmospheric pressure). Indeed, the lattice constant of GaP at atmospheric pressure is slightly smaller than that of GaAs.

It is of interest to note that behaviour of the band gaps at Γ and X illustrated in Fig. 10.10 for GaAs (i.e. a large positive pressure coefficient for the Γ gap and a small negative one for the indirect (X) gap) represents a general trend that applies to all diamond and zinc-blende semiconductor crystals.

Let us now consider a GaAs–$Ga_{1-x}Al_xAs$ quantum well structure. The

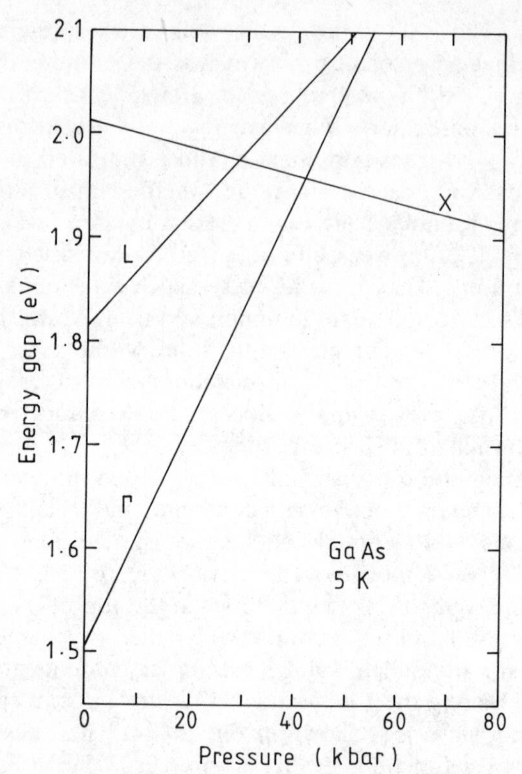

Fig. 10.10. The variation of the energy separation (band gap) between the top of the valence band and the primary (Γ) and secondary (X, L) conduction band minima of bulk GaAs as a function of hydrostatic pressure in kilobars (kbar). (1 bar = 1 Torr.)

lowest conduction band level is now at a somewhat higher energy (above the bulk conduction band edge of GaAs) because of the confinement effect. Similarly, the uppermost valence state lies below the bulk valence band edge of GaAs, and the band gap of the quantum well or superlattice structure is a little larger than the band gap of GaAs. We can shine laser light of energy higher than the band gap upon a sample containing the superlattice. This excites some valence electrons from the valence band high into the conduction band. These electrons rapidly dispose of their kinetic energy by emitting phonons and relax down to the bottom conduction level of the superlattice. There they form excitons with the holes left behind in the valence band. The exciton binding energy is only a few millielectron volts. We considered such excitons in more detail in Chapter 7. After about 1 ns the exciton dissociates and the electron jumps into the empty slot in the valence band. The excess energy is emitted as a photon. This exciton luminescence is recorded by a detector. It gives a strong characteristic peak at an energy above the bulk band gap energy of GaAs. This transition energy is shown in Fig. 10.11. It corresponds to the zero-pressure point on the second curve from the bottom. The lowest point, at about 1.52 eV, is the fundamental band gap of bulk GaAs. This is approximately the energy we would obtain in an analogous experiment with a sample of bulk GaAs.

If we now apply hydrostatic pressure, we can follow the luminescence transition as a function of pressure until we reach the crossover point where the lowest confined level hits the lowest *X* valley and moves above

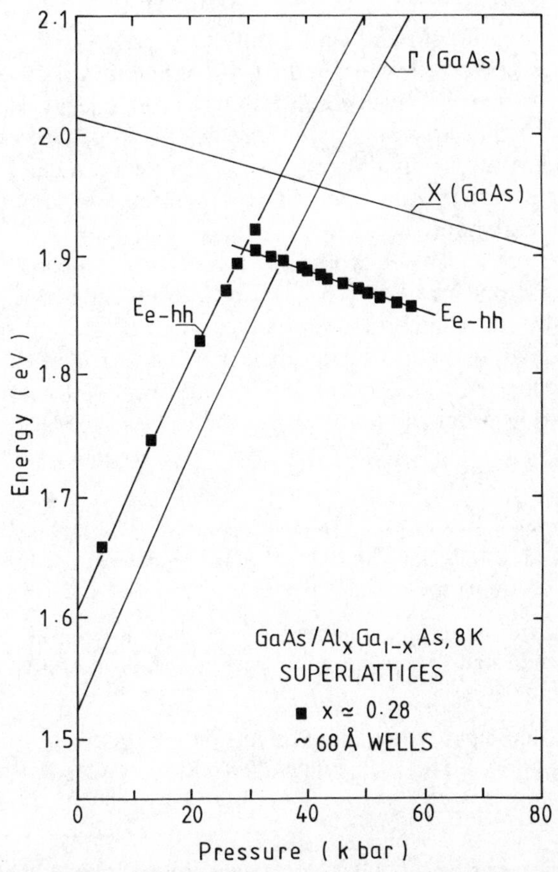

Fig. 10.11. Experimental results (solid squares), showing the pressure dependence of the optical transition (luminescence) across the forbidden gap of a GaAs–$Ga_{0.72}Al_{0.28}As$ superlattice at low temperatures. The wells and barriers are about 68 Å thick; *e–hh* indicates that an electron and a heavy hole are involved in the exciton recombination that gives rise to the signal.

it. Above the crossover the lowest conduction band level is X-like. However, we know that the optical transition across the indirect gap is very weak compared to the direct gap transition. It follows that the pressure at which the crossover occurs is clearly identifiable from the luminescence spectra, since the intensity of the signal rapidly drops at that point. Also, the pressure coefficient of the X point is very different from that of the Γ point and the energy of the weak signal obtained after the crossover actually decreases with increasing pressure.

Once the crossover point has been identified, we can compute the valence band offset ΔE_v from

$$\Delta E_v = E_g^X + |E'_{\text{ex}}| - E_{e-hh}(\text{crossover}) - \Delta_{hh}, \tag{10.6}$$

where E_g^X is the indirect band gap of bulk $Ga_{1-x}Al_xAs$, E'_{ex} is the exciton binding energy corrected for the effect of confinement (about 6–9 meV), and E_{e-hh}(crossover) is the observed transition energy shown in Fig. 10.11, taken at the crossover point. Δ_{hh} is the (positive) energy of the confined level at the valence (heavy hole) band edge, measured from the bottom of the well in question. These energies are familiar from the account of confinement effects in Chapter 7.

The pressure at which the crossover occurs can be measured with great accuracy. The input into eqn (10.7) therefore depends on empirical data that are known with uncertainties of the order of several millielectron-volts. Consequently, in this procedure, the final error in the determination of the valence band offset is significantly reduced compared to the error expected in other methods.

Problems

10.1. Draw a rough graph of the arrangement of atoms at the ideal (001) surface of a single crystal of (1) NaCl and (2) Si. Find the nearest-neighbour separation between atoms forming the surface layer in each case (a(NaCl) = 5.64 Å.)

10.2. Calculate the wavelength λ and the energy E of an electron beam chosen so that λ equals the smallest separation between atoms at the (001) surface of (1) NaCl and (2) Si.

10.3. Consider an impurity (e.g. magnesium) on the surface of a crystal studied in the reactor shown in Fig. 10.1. Predict the velocity of the electrons emitted in the Auger process described in Fig. 10.6. (Take $E_{\text{K}} = -80$ eV, $E_{\text{L}} = -20$ eV and $E_{\text{M}} = -7$ eV)

10.4. Consider a sample of bulk germanium doped with 10^{16} donors cm^{-3}. Determine the smallest value of the magnetic field at which a cyclotron resonance experiment can be successfully performed at temperature $T = 10$ K.

10.5. Use the experimental result shown in Fig. 10.11 to determine the magnitude of the valence band offset ΔE_v in the structure in question.

11

Application of semiconductor microstructures in electronic devices

11.1. Planar silicon technology. MOSFET and bipolar transistors

Transistors are the basic building units of electronic circuits. It is, therefore, natural that advances in the physics of very small semiconductor structures, and in semiconductor technology in general, have been rapidly translated into progress in transistor design and manufacture.

The first transistors were made of germanium. However, in the late 1950s, silicon, with its larger band gap and better mechanical and thermal properties, replaced germanium as the most favoured material for transistor manufacture. A concerted effort over several decades has produced a sophisticated technology based on high-purity crystalline silicon. The dominating influence of silicon technology in electronics industry can hardly be over-estimated. The rate at which transistors and circuits in general have been miniaturized is quite remarkable. While the smallest feature of a circuit was about 30 μm in 1960, by 1985 it had been reduced to about 1 μm. A single chip can support a million transistors capable of carrying out high-speed logic operations. It is at this level of sophistication that the physics of semiconductor microstructures described in this course becomes relevant.

Two kinds of transistors have been commonly used in the electronics industry: the bipolar transistor and the field-effect transistor (FET). Bipolar transistors are often used as basic building blocks of a computer's central processing unit. This is the part of the machine that does data processing and operations. FETs are used mainly to build computer memory. This division stems from the fact that bipolar transistors have been faster in switching than FETs. However, in smaller computers, the FET has been used in both functions and, as circuit design changes with changing technology, this trend is likely to continue.

Any transistor can be viewed as a switch. By applying an electric field to the base, the current is switched on and flows from one electrode (emitter) to another (collector). The field can be turned off to stop the

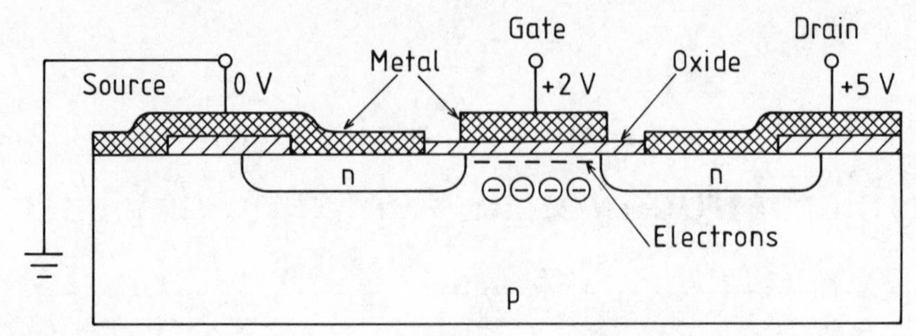

Fig. 11.1. A schematic representation of the layout of a planar silicon n-channel MOSFET.

flow. When referring to an FET, the elements corresponding in their function to base, emitter, and collector are known as the gate, source, and drain, respectively. A typical layout for a planar silicon FET, emphasizing the material aspects of the design, is reproduced in Fig. 11.1. The active part of this device consists of an MOS structure as described in Chapter 6.

When a positive voltage is applied to the metal gate of this n-channel FET, electrons located in the substrate farther away from the semiconductor–oxide interface (open circles with the minus sign) are attracted towards the channel region near the interface (the interrupted line) and the channel conducts. The current is collected in the drain region by the voltage between source and drain electrodes. When the gate voltage is reduced, electrons diffuse away from the channel and the current stops. The speed of the device depends on the speed with which electrons can be transferred across the gate region. The curves near metal–semiconductor contacts indicate the space charge regions that determine the potential profiles at the end points of the active channel. The shorter the channel, the more important these end point regions become.

A similar planar layout is used for bipolar transistors. The band diagram describing the functioning of these devices is shown in Fig. 11.2. The structure consists of three sections, which are shown separately in Fig. 11.2a. The first section, on the left, is heavily n-doped. The second and the third sections are moderately p- and n-doped. When these components are joined together to form a device, we have two p–n junctions in a back-to-back configuration, shown in Fig. 11.2b. At equilibrium, the Fermi level of the system must be the same everywhere, as indicated. The n–p emitter barrier is higher than the barrier separating the base and the collector, as expected from the above-mentioned doping levels. If a forward bias voltage is applied, the emitter barrier is reduced

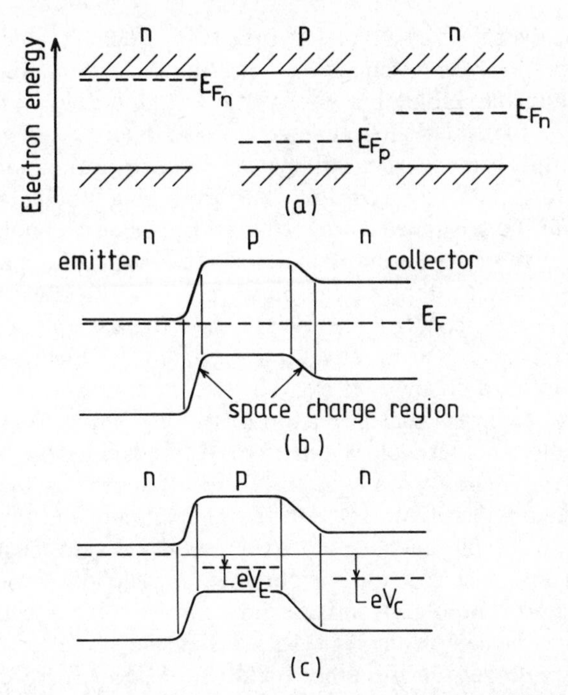

Fig. 11.2. The band diagrams of a n–p–n bipolar transistor (BT), in three stages. (a) The band diagram of the three components of the BT taken independently. (b) The band diagram of the BT components joined together (no bias). (c) The BT biased in the active position. V_E is the forward-bias voltage of the emitter. V_c is the reverse-bias voltage of the collector.

and electrons can flow into the base region (Fig. 11.2c). The collector is reverse-biased. This barrier prevents holes from entering the n-type collector and collects the injected electrons from the base region.

The gain achieved in transferring injected carriers from the emitter to collector depends on the efficiency of this transfer. This efficiency can be significantly reduced by collisions with phonons (lattice vibrational waves), by trapping and recombination at impurities, and because of the possibility of electrons escaping from the channel. The speed of the carrier transfer, and consequently the switching speed, depends on the speed with which the carrier moves across the base region and on the base width.

11.2. Scaling of planar silicon devices

The planar configuration illustrated in Fig. 11.1 is a natural choice given the character of the manufacturing process, originating in a flat high-quality substrate and involving processing such as polishing, layer deposition and doping, etching, pattern making, and coating. It is, therefore, understandable that the attitude adopted by designers and technologists is that the planar character of the device must be preserved so long as improvements can still be achieved by other means. Also, the huge investment required for the developing new materials and structures, and the advanced state of the art of silicon technology, dictate that new circuitry be based on planar silicon devices as long as possible.

The simplest way to improve the performance and the packing density of silicon FETs is to reduce their overall size. Ideally, one would like to do this without having to modify the circuit design. Such modifications might prove to be a costly exercise. This means that one would like the smaller device be just a scaled version of the old, larger one. However, by altering the geometrical layout, i.e. by reducing the size of the individual components, we are imposing new constraints upon external parameters such as applied voltages. If, for example, the functioning of the device is to remain unchanged, the external voltages must be altered so that the same electric fields are obtained at junctions as in the larger original structure. Bipolar transistors are not suitable for such scaling because, as can be seen from Fig. 11.2, scaling the size alters the turn-on emitter–base voltage. On the other hand, the MOS structure that forms the basis of the planar silicon FET (which is why this device is often referred to as a MOSFET) is more amenable to straightforward scaling.

The voltages required in the operation of a MOSFET are linked to the size of the active part of the device in a manner that is apparent from Fig. 11.3. On the left we see a simplified diagram of the device pictured

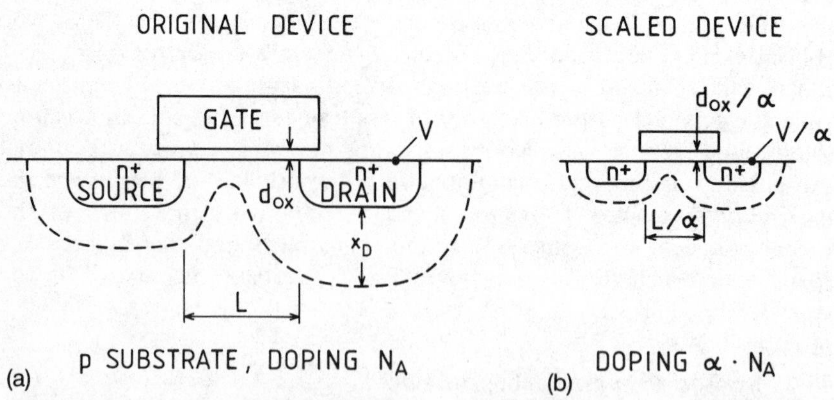

Fig. 11.3. (a) A highly schematic diagram of a MOSFET. (b) Scaled MOSFET: α is the scaling factor. This scaling preserves the magnitude of the electric field of the unscaled device.

in Fig. 11.1, and on the right a device whose dimensions are scaled by a factor α. The basic principle of scaling, to reduce if possible all dimensions of the device by a common factor, implies that the external voltages are scaled accordingly to preserve the electric fields. Under such circumstances, the depletion region remains approximately the same in both structures. This means, however, that the depletion region takes up a larger fraction of the device. Eventually the depletion regions originating from the two electrodes must merge and destroy the operation of the device. Thus, an important task for the designer is to reduce the depletion region that surrounds the source and drain electrodes and extends into the substrate (broken line in the figure). Furthermore, when a voltage is applied to the gate to turn the MOSFET on, a small depletion region is formed under the gate. The basic function of the gate is to make it possible to reduce the barrier between the substrate and source in the region close to the gate so as to allow the current to flow from source to drain. This occurs at a critical gate voltage which must be well defined and consequently substantial in magnitude. Therefore, the effective channel length that determines the field gradient cannot be allowed to deteriorate too much by the overlap between the two depletion regions in the scaled structure.

For an abrupt junction, the width of the depletion layer is proportional to V/N, where N is the dopant concentration and V is the sum of the applied substrate voltage, the junction potential, and the applied drain–source voltage. Hence, in the scaled device, the depletion width is scaled down by an increase in the doping concentration and by a reduction of the voltages, both by α. The scaled gate voltage in fact gives the same electric field as in the original structure because of the scaled insulator layer at the gate interface. The charge density in the inversion layer is the same. The current of the scaled device is therefore also largely unchanged. Since the carriers move laterally, and the lateral field remains the same, their lateral velocity is not altered. With the gate voltage scaled by α and current unchanged, the gain in the scaled system is enhanced by α. Finally, reduction in voltage implies reduction in power dissipation, provided that only the channel length L is scaled and the device width remains constant.

The approach to scaling outlined above was motivated by the desire to reduce the size of the structure while retaining the same electric field and normal functioning of the device. Another way to look at the problem of scaling is to require that the power supply voltage is not altered. This leads to larger electric fields in the smaller device. In such fields, electrons acquire large kinetic energies and some effort must be made to contain unwanted hot-electron effects. These effects impose a practical limit of the applicability of this approach.

11.3. High-electron-mobility transistor (HEMT)

Another way to increase the transistor speed is to abandon silicon technology and choose a different material with more favourable band structure parameters. This approach was taken up in the late 1960s as an alternative to be developed in parallel with the mainstream silicon effort, when it became apparent that, at least for some applications in optoelectronics, a material with larger (direct) forbidden gap and smaller effective mass than that offered by silicon was indispensable.

In our account of the electronic structure of semiconductors in Chapter 5, we noted that there is a significant variation in the electron effective mass. In particular, the electron mass is about five times smaller in GaAs than in silicon. The electron mobility is inversely proportional to the effective mass, so that the smaller mass of conduction electrons in GaAs offers an opportunity to improve significantly the transistor switching speed. However, it was not until after the advent of the modulation-doped structures, described in Chapter 8, that a full-hearted effort was made in the early 1980s to exploit GaAs in transistor circuitry. At low temperatures, when lattice vibrations are suppressed, the main source of electron collisions (and consequently the main cause of reduction in the electron flow) is the presence of impurities. Since dopants supply carriers without which a semiconductor material cannot be used in electronic devices, their presence in the material is indispensable in semiconductor technology. The essence of the modulation-doping principle is to separate the electron current in the active channel from the region where the dopant impurities are located. This is achieved by employing heterojunctions that confine carriers on the side of the interface with deeper energy levels. The dopant impurity atoms are located on the other side of the heterojunction.

The simplest heterojunction structure that can be operated as a switch and strongly resembles a silicon MOSFET is shown in Fig. 11.4. The thin undoped alloy layer acquires the role of the oxide insulator in MOS structures. It also confines the electrons, supplied by the impurities located in the thick intentionally doped alloy layer, in the undoped GaAs layer. The near-triangular confining barrier is shown in Fig. 11.5. The confined electrons at the interface on the GaAs side occupy the lowest sub-band and only about 2% of their wave function in the direction perpendicular to the interface penetrates into the barrier material. The position of the sub-bands (energy levels E_1 and E_2), and the approximate shape of the electron wave functions associated with these sub-bands in the direction perpendicular to the interface, are illustrated in the lower diagram. A notable contribution to the height of the confining barrier is made by the conduction band offset, although the

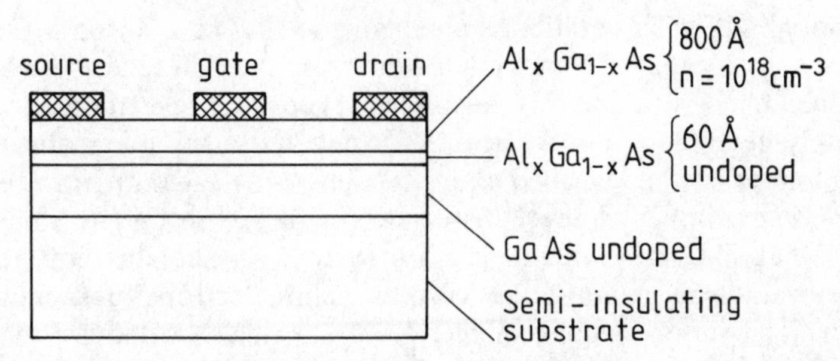

Fig. 11.4. A block diagram of a HEMT. The active electron channel is in the undoped thin layer of GaAs. Electrons move along the interface as in a MOSFET.

barrier also depends on the space charge potential and consequently reflects the doping concentration. The length of the active channel (roughly equal to the gate width) parallel to the interface is typically about 1 μm. This device is often called high-electron-mobility transistor (HEMT).

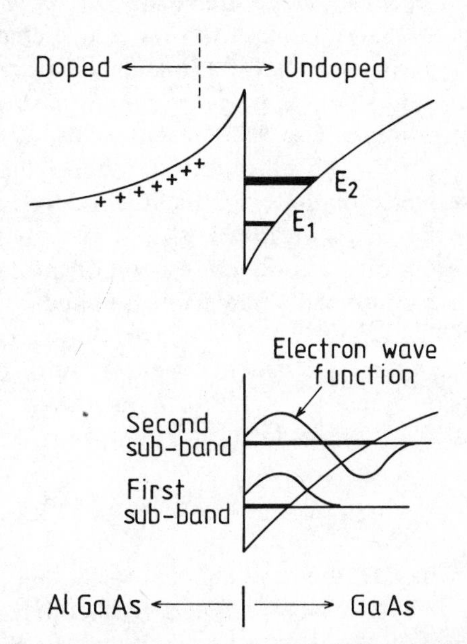

Fig. 11.5. The sub-band structure at the interface of the GaAs active channel in a HEMT structure (Fig. 11.4). E_1 and E_2 are the confined levels. The approximate positions of E_1 and E_2 as well as the shape of the wave functions are indicated in the lower part of the diagram. In the upper part, an approximate form of the potential profile is shown, including contributions of the conduction band offset and of the space charge potential.

Although HEMT operation is analogous to that of a silicon MOSFET, there are significant differences that stem from the difference between the electronic band structure of silicon and GaAs. These differences can ultimately be exploited in the search for new more suitable materials.

In Chapter 8, we outlined the effect of (hot) electron transfer into higher-lying conduction levels associated with the secondary minima at the so-called X point. Since the electron velocity is actually reduced upon such elevation in energy, the current exhibits strong non-ohmic behaviour. The onset of these effects occurs at fields of about $3\,\mathrm{kV\,cm^{-1}}$. This critical value of the field, which is obviously a key limiting factor, depends on the energy separation between the lowest (Γ) and secondary (X and L) valleys. It also means that the threshold for non-ohmic behaviour could be increased if a material in which this separation is larger were to be used instead of GaAs. For example, in InP this energy separation is 0.6 eV, as compared to about 0.3 eV in GaAs. Such a material offers a useful alternative to GaAs, since higher drift velocity and velocity overshoot can be achieved.

Another important parameter peculiar to the HEMT structure is the barrier height. Hot electrons have high kinetic energy and they can be excited over the barrier and be lost from the active channel. Hence, it is desirable to have a material with a higher barrier. Since the most important contribution to barrier height comes from the band offset, the aluminium concentration is usually chosen to be as high as possible (about 38%; above this concentration the alloy band gap becomes indirect). The effective confining height is also reduced by the magnitude of the sub-band energy E_1, shown in Fig. 11.5 (i.e. by the height of this level above the bottom of the conduction band of bulk GaAs).

An approximate formula obtained from detailed calculations can be used to estimate E_1 for a given sheet concentration N_s of the electron gas in GaAs. In good samples, the sheet concentration of ionized impurities whose presence might affect the value of E_1 is at least ten times smaller than N_s and can as a first approximation be ignored. We have

$$E_1 = \left(\frac{3}{2}\right)^{5/3}\left(\frac{e^2\hbar}{\epsilon_0\epsilon m^{*1/2}}\right)^{2/3} \times 1.2N_s^{2/3}. \tag{11.1}$$

The electron sheet density N_s in the channel in the high doping limit can be obtained from the Gauss law applied to the heterojunction in the manner familiar from its application to p–n junctions. Then N_s is roughly

$$N_s \simeq \frac{\epsilon_0\epsilon}{e^2 d}[\Delta E_c - (E_c - E_F)], \tag{11.2}$$

where d is the thickness of the thin undoped (spacer) $Ga_{1-x}Al_xAs$ layer shown in Fig. 11.4; $E_c - E_F$ is the difference between the conduction band edge of $Ga_{1-x}Al_xAs$ and the Fermi energy; and ΔE_c is the conduction band offset. $\Delta E_c \simeq 0.3$ eV and $E_F - E_c \simeq 0.1$ eV, so that d must be small ($\simeq$100 Å) for N_S to be large. N_s is typically 10^{11}–10^{12} cm^{-2}. However, d must be large enough to protect the conduction electrons in the channel from interactions with the long-range Coulomb potential of the impurities in the doped (thick) $Ga_{1-x}Al_xAs$ layer.

Scaling HEMT is a very much the same process as scaling MOSFET because of the geometrical similarities. Hence the objectives of scaling—reduction of the area occupied by the device on the chip and shortening the length of the active channel to increase the switching speed—also remain. We can estimate the switching speed as a function of channel length by the simplest means. Using mobility $\mu = 10^5$ cm^2 V^{-1} s^{-1} ($\mu = e\tau/m^*$) and $m^*/m = 0.067$, we obtain a mean free time (average time between subsequent collisions) of 3.8 ps, which is greater than the transit time in a short-channel HEMT (if we take velocity $v = 5 \times 10^7$ cm s^{-1}, this transit time implies channel length in excess of 10^{-4} cm).

The large values of mobility (>10^5 cm^2 V^{-1} s^{-1}) achieved at low temperature in modulation-doped structures with long channels may have little meaning in very small devices. In a short-channel device, the potential profile determined by the spatial form of the depletion layer and by the actual layout of the device components, material quality, etc., must be fully accounted for in the assessment of the operational switching speed. We can estimate the mobilities achieved in a short-channel HEMT by inferring them from measured resistance. It turns out that μ is of order 30 000 cm^2 V^{-1} s^{-1}. However, this mobility is still much larger than $\mu \simeq 600$ cm^2 V^{-1} s^{-1} achieved in silicon MOSFETs.

11.4. Hot-electron transistors

Just as GaAs heterojunction structures are used to replace silicon MOSFETs by a faster planar FET-like transistor, so this idea has been used to make a GaAs-based bipolar transistor. The band diagram for this device is formally quite similar to that we drew above to describe the silicon-based device. GaAs is used to make the base, and the $Ga_{1-x}Al_xAs$ alloy material serves as emitter and collector. However, the performance of the GaAs heterojunction bipolar transistor is not much superior to the best silicon bipolars, and efforts were made to look for an alternative fast-switching GaAs-based device. This effort led to the discovery of the so-called hot-electron transistor.

The band diagram explaining the functioning and the layout of this device is shown in Fig. 11.6. It is a vertical electron transport device, unlike the MOSFET or HEMT in which carriers in the active channel

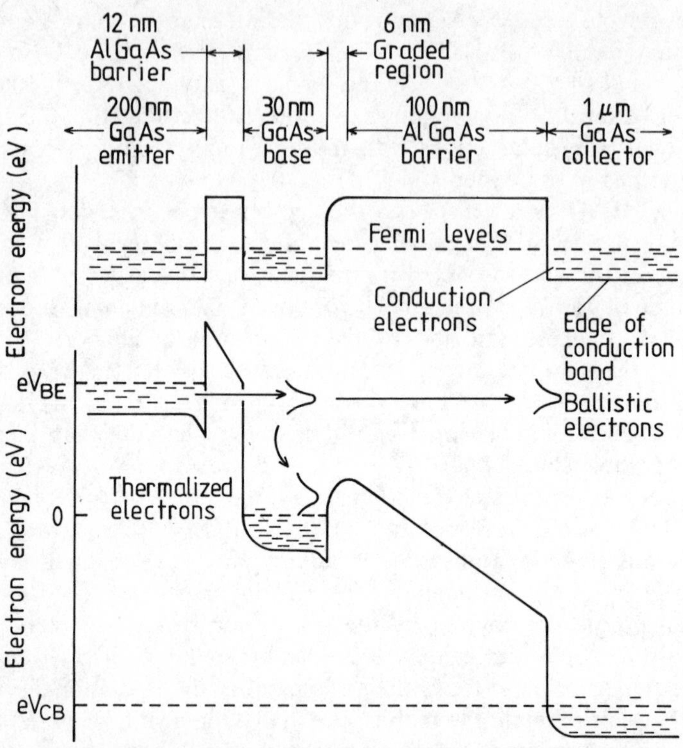

Fig. 11.6. A block band diagram of a hot-electron transistor. The lower part shows the potential energy profile with the bias voltages included.

move in the direction parallel to the interface. The upper part of Fig. 11.6 shows the arrangement and widths of the individual semiconductor layers, and the approximate profile of the conduction band. The step-like character of this potential profile is given by the familiar conduction band offsets, and to some extent by the space charge contribution. When an external voltage is applied (the bottom part of the figure), electrons are injected perpendicularly to the interface plane into the GaAs base. They tunnel through the narrow barrier separating the emitter and base, lowered by the applied field. As the diagram suggests, these injected electrons are hot, i.e. they have substantial kinetic energy thanks to the voltage V_{BE} between the emitter and base.

Since the base is narrow, and is formed by high-purity undoped GaAs, most of these electrons can get across the base without collision. This means that these electrons can reach the limiting velocity described in Chapter 8. This is illustrated in Fig. 11.7. The electrons that get across without collision are called ballistic electrons. The transit time across the

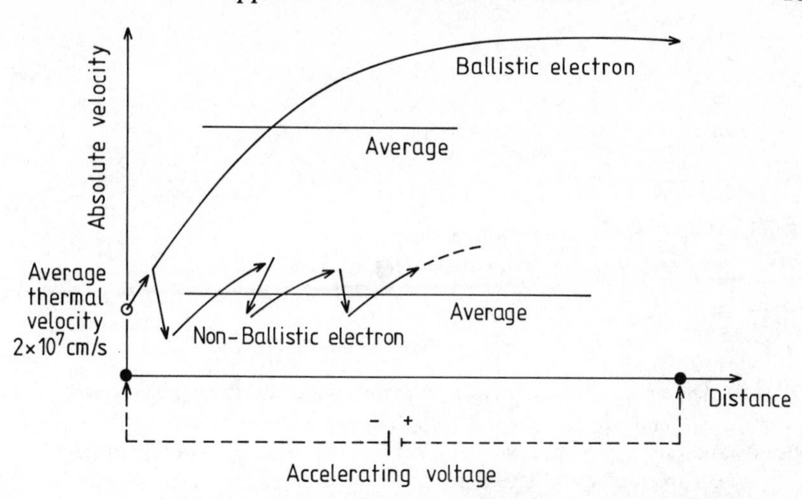

Fig. 11.7. The velocity of ballistic and normal (thermal) electrons in the base of a hot-electron transistor as a function of position from the entry point.

base in this device is about 30 fs (femtoseconds). This implies a frequency response limit for this device of 5 THz. However, for the current to complete the cycle, the carriers must cross the (much wider) collector layer, so that the response time is longer. In practice, the actual response is limited by the time needed to charge up the capacitances given by the width of the emitter–base tunnelling barrier and the base–collector barrier, and by the corresponding voltage drops. A realistic estimate of the characteristic transit time of this device is of the order of 1 ps.

The criteria for performance improvements of the hot-electron transistor are quite predictable from our earlier considerations concerning the HEMT. We can reduce the transit time by making the device smaller. In particular, we might endeavour to alter the thickness of the layers so that the parasitic capacitance determining the charging time is also diminished. To increase the current, we must decrease the width of the emitter–base barrier. Higher band offset is also an advantage for better control of tunnelling and hot-electron effects.

We have seen that the functioning of transistors whose size is constantly being reduced is threatened unless we succeed in shrinking the depletion layer widths. This can only be achieved by higher doping densities, which have a number of undesirable effects (e.g. high electric fields, the increased possibility of atomic diffusion, and impurity complex formation). One way to avoid high doping densities is to replace the semiconductor base by a metallic one. A schematic diagram of a metal-base transistor (MBT) is shown in Fig. 11.8. The MBT is also a

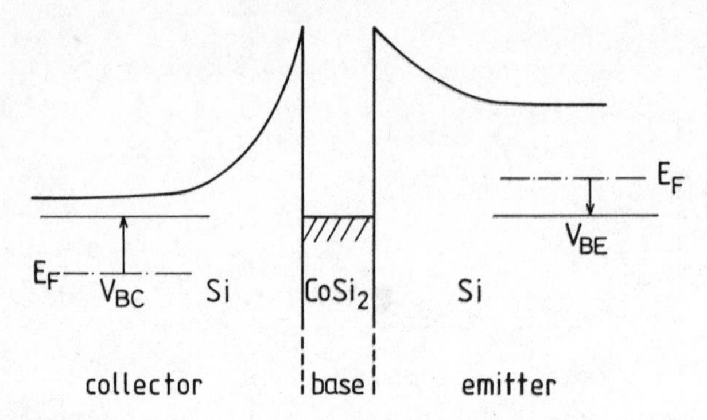

Fig. 11.8. The energy band diagram of a metal-base transistor. The base is made of $CoSi_2$. V_{BC} and V_{BE} indicate the applied voltages between the base and the collector and emitter, respectively. E_F is the Fermi energy. The upper part of the occupied electron levels in the metal is indicated by hatching.

vertical device, exploiting the possibility of fast ballistic transport across a narrow base. The functioning of this structure in Fig. 11.8 can easily be understood if we realize that we have here two Schottky barriers (described in detail in Chapter 6) in a back-to-back arrangement.

The first MBTs were made with silicon emitters and collectors. The electrons are injected into the metallic ($CoSi_2$) base by thermionic emission from the forward-biased Schottky (emitter) diode. A significant fraction of these carriers cross the base ballistically because the base thickness is smaller ($\simeq$100 Å) than the mean free path of electrons in the metal. This means that electrons reach the collector at a constant energy above the metal free-electron Fermi energy. They are collected by a reverse-biased collector diode, as indicated in the figure. Measurements show that in the 'ballistic' regime the mean free path is about the same as the 'normal' mean free path in metals like gold or silver (a few hundred angstroms). Some losses occur in the collector junction. The electrons entering the collector experience a strong electric field owing to the sudden change of potential at the metal–semiconductor boundary. This leads to electron reflections from the interface.

The base of the silicon device is made of a silicide, $CoSi_2$, instead of any of the conventional metals. This is because the lattice constant of $CoSi_2$ is very similar to that of silicon and it is possible to grow layers with sharp, high-quality interfaces. The mean free path in $CoSi_2$ is only slightly shorter than that in other metals. However, because of the reflections at the interface, the actual (observed) mean free path l depends on the width of the metal layer. For example, at low temperatures $l = 550$ Å in a structure in which the base is 600 Å thick, but it is only 200 Å when the base width is reduced to 50 Å. In spite of the near-perfect lattice matching, the practical difficulties in making a good-quality metallic epitaxial layer represent the most significant obstacle in achieving high-performance MBTs.

The possibility of using a metal base has also been considered in connection with GaAs technology. The problems encountered here are again related to the quality of the layers and interfaces.

The high current density in vertical electron transport devices such as the MBT or hot-electron transistor greatly increases demands on the degree of perfection of the deposited layers. For example, imperfect growth of the $CoSi_2$ metallic base layer may lead to clusters of cobalt material. Such clusters exhibit lower resistivity. This leads to uneven heating of the base and to current fluctuations.

In a GaAs hot-electron transistor, the barriers are made of $Ga_{1-x}Al_xAs$ alloys. The lattice of these alloys can be visualized as that of a perfect GaAs with gallium atoms replaced randomly by aluminium so as to ensure that on average there are x atoms of aluminium for every $1-x$ atoms of gallium. In practice, the distribution of these atoms may not be strictly random. Then—depending on the quality of the epitaxial growth—certain areas contain either gallium-rich or aluminium-rich material. Such irregularities may also be introduced by the presence of lattice defects that weaken the bonds between atoms in the lattice and disturb the growth.

As a result, the electrons injected in the direction perpendicular to the interface experience different resistivities depending on the spatial location of their entry. In fact, in the case of the GaAs–$Ga_{1-x}Al_xAs$ heterojunction we can visualize this effect simply as a fluctuation in the barrier height. For instance, a local reduction in aluminium concentration means that the actual conduction band offset at that point is smaller. Hence, some areas are heated more than others. The effect of such heating is to excite localized atomic vibrations that may 'kick' an atom out of its normal lattice position and force it to wander through the lattice. The heating therefore tends to 'burn holes' in the material. The electrical properties in the vicinity of these 'holes' deteriorate even further and so does the device performance.

As the cross-sectional area of the device decreases with miniaturization, the current density reaches values previously unheard of. Consider, for example, a current of 10 mA, which is quite common in transistor circuitry, carried by a device of cross-sectional area 10^{-8} cm^2! Given the danger of electromigration of excited atoms mentioned above, such large concentration of energy in a small structure requires that the device be efficiently cooled. Let us estimate the role of energy dissipation in material considerations. The power dissipated in a logic chip operating at

Table 11.1. Thermoconductivity ν

Material	$\nu(\mathrm{W\,K^{-1}\,cm^{-1}})$
Si	1.5
Ge	0.7
GaAs	0.46
InP	0.68
SiO_2	0.14
CaF_2	0.11

frequency f is given approximately as

$$P = n_g f w_g, \tag{11.3}$$

where n_g is the density of gates on the chip and w_g is the energy required to switch the gate on and off (i.e. per cycle). It is convenient to express w_g as a product of some mean delay time τ_D and the corresponding power delay P_D

$$w_g = 2P_D\tau_D. \tag{11.4}$$

If we choose constant P/n_g, then w_g is inversely proportional to frequency. Since the operational frequency f is limited by the achievable value of τ_D, we can see that power considerations strongly affect the frequency characteristics of the device.

The rate of flow of heat in a solid is measured by the thermoconductivity ν. The value of ν for a few important semiconductors is given in Table 11.1. Since P must depend on the threshold voltage V_{th} needed to switch the gate on—which is given by the choice of materials and structure used to make the device—it is the best combination of V_{th} and ν we must be looking for. In semiconductor devices, V_{th} is in the range of 0.5–1.5 eV. As far as ν is concerned, Table 11.1 shows that silicon is superior to GaAs and InP.

11.5. Physical limits of integrated circuits

The circuits in which the characteristic number of transistors on a single chip is of order 10 000 or larger are called very large-scale integrated circuits (VLSI). Another, and no less necessary, ingredient in the definition of VLSI is the smallest feature size. As Table 11.2 shows, the feature size scales with the packing density and the number of devices actually assembled on a single chip. The smallest feature size characteris-

Table 11.2. Changes in miniaturization of transistor technology

Integration scale	Number of transistors per chip	Smallest feature size (μm)	Era of development
Small (SSI)	1–100	10	1960's
Medium (MSI)	100–1000	5	1970's
Large (LSI)	1000–10 000	3–1	1980's
Very large (VLSI)	>10 000	<1	1980's
Ultra high (UHSI)		<0.3	

tic of VLSI is of order 1 μm. It is a measure of the level of the overall advance in technology because it reflects the state of the art in materials growth and processing, which imposes the ultimate limits on what can be achieved. The transistor is a basic unit product of this technology and a key building block of circuitry. Consequently, when we consider physical limits of semiconductor circuits, we must consider first the physical limits on the size of a transistor.

There are a number of factors that may be thought to limit the size of semiconductor devices, such as tunnelling between layers, device–device interactions, and a variety of quantum interference phenomena associated with the microscopic properties of atoms and electrons in the crystalline lattice. However, all these effects are associated with a characteristic limiting length that is much too small. The decisive limiting factor stems from the very nature of conduction in semiconductors that form the body of any transistor.

In general, all transistors consist, totally or in part, of doped semiconductor layers. Therefore, the need to provide free carriers in a given volume is the most fundamental condition on which the functioning of any transistor depends. The limiting size is given by the smallest (theoretical) volume that can still support the largest tolerable and achievable concentration of impurities. It turns out that this minimum volume is a cube of about 400 lattice constants on a side (i.e., about 2000 Å in length). Since a transistor consists of three distinctly different parts (source, gate and drain), each requiring different doping levels, the minimum length reserved for one transistor must be three times as long. Hence, the smallest transistor would take up an area on the chip of about 10^{-8} cm^2, or about 1 μm on a side.

The number of transistors we can accommodate on a chip also depends on how large a slice of semiconductor can be made without deterioration in material quality or production efficiency. The largest chips available are made of silicon. They are made by slicing cylindrical ingots of single-crystal silicon into thin wafers. Each wafer is divided into a large number (perhaps 100 or so) chips. The chips are tested, and those containing defects are simply thrown out, so that it is uneconomical to make large chips unless uniform quality can be maintained over the whole wafer. Let us assume, for the sake of simplicity, that we have produced a chip whose area is $10\,\mathrm{cm}^2$. Only about 10% ($1\,\mathrm{cm}^2$) can be used for transistors. The remaining area is taken up by interconnections. Hence, with each transistor occupying $10^{-8}\,\mathrm{cm}^2$, there is room for 10–100 million transistors. The largest number achieved, for example, in 1986 was 2 million on a somewhat smaller chip. There is, therefore, an increase of about 10 to achieve before the physical limit predicted above is reached.

The limiting size suggests that there is still a substantial scope for improvement beyond VLSI. The first question to ask must be about the ways this physical limit could be achieved. The concepts of heterojunction microstructures introduced in Chapters 6–8, and the microdevices that are based on these concepts, such the scaled MOSFET, HEMT, and the hot-electron transistor, go a long way towards this goal. We can then start wondering about the means of going beyond this limit.

For ultrafast switching, silicon is inferior to compound semiconductors such as GaAs. Since the main driving force at the advanced state of miniaturization is speed, the compound semiconductors that have catered mainly for specialized applications might be seen as a natural replacement for silicon. However, it is worth remembering that chips of GaAs or InP have always been smaller and less well-characterized, so that the progress towards higher integration is slower and unlikely to reach the theoretical limit, expected to be about 10 million transistors per chip.

In arriving at the above limit of semiconductor technology, we have assumed that the scaling procedure can successfully be extended as far as this limit. It remains to be seen to what extent this assumption is realistic. We also assumed that the parasitic capacitances and problems with interconnections can be taken care of. In fact when it comes to the cost, reduction per transistor is less important than the cost reduction per interconnection. This is easy to understand, because interconnections occupy much of the volume of integrated circuitry. That means that there is a scope for improvement, should it prove possible to increase the percentage of space taken up by the active devices. It is important to reiterate that the limiting point is reached not because of some quantum-mechanical constraint associated with the microscopic nature of atoms and their proximity across the interfaces, but because of the nature of the transistor mechanism. Hence, we can conclude that we might be able to stretch the limit further provided one of the several assumptions we have made in our assessment here proves to be too restrictive.

In Table 11.2 the VLSI technology is followed by UHSI, in which the

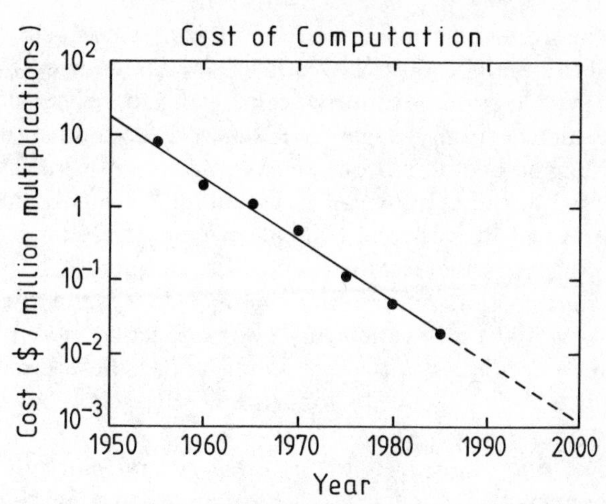

Fig. 11.9. The cost of computations in US dollars per million operations at various periods.

smallest feature size is of order 0.1 μm or less. This proposition is by no means fanciful and is based on shrewd assessments of the future potential of semiconductor microstructures. The success of the electronics industry has been maintained not only by the improvement of performance with time that is apparent from Table 11.2, but also by the logarithmic reduction of cost, both in terms of cost per operation (Fig. 11.9) and per bit of storage (Fig. 11.10). It follows that, unless improvements

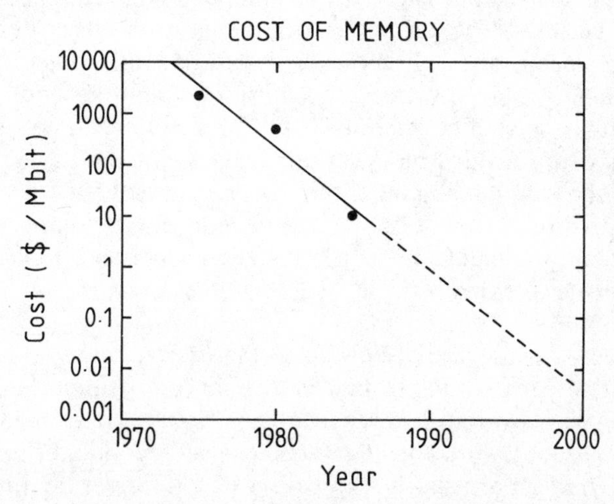

Fig. 11.10. The cost of memory in US dollars per megabit at various periods.

approaching the extrapolations indicated in these figures actually materialize, the industry might face a major crisis. Of course, the cost of computers in the twenty-first century is not going to be 'negative' as the extrapolation, taken literally, suggests. However, the momentum may be kept by providing an overall performance improvement while the cost per bit remains largely static. Information technology, which is unthinkable without electronics and without the modern materials science on whose success the progress in electronics depends, is rapidly becoming the largest industrial block in the world economy. Hence, the consequences of such a crisis would reach practically every aspect of modern life. The pressing need for progress is overwhelming under these circumstances, even though the cost of going beyond the VLSI technology may seem prohibitively large.

For example, one of the candidates for a central place in the future UHSI technology is the concept of quasi-three-dimensional, quantum dot resonant tunnelling device structure. Such a structure evolves naturally from the established planar silicon and GaAs technology in that its building block is a quantum dot (or box) created by extending the confinement effect from one to three dimensions. Carriers are then confined into small boxes of dimensions not exceeding a few hundred ångstroms in length. The transport process is initiated by an external voltage pulse that can line up the position in energy of the confined electron levels in the adjacent boxes. These levels can be modelled by methods described in Chapter 7. The current tunnels from one box to another depending on the geometrical and electrical parameters (see Fig. 11.11). Layer upon layer of these quantum dot systems may be grown to achieve a three-dimensional 'chip'. There are a number of fundamental decisions that must be made at the very beginning of such a development programme. For instance, if this scheme is to be implemented within the silicon technology, a new insulator must be found to replace silicon dioxide with a crystalline material well lattice-matched to silicon and suitable for epitaxial growth. Various candidates have been considered (e.g. CaF_2 and sapphire). The physics of interfaces between silicon and these insulators offers a number of new phenomena. Similar fundamental problems arise in connection with transport properties of electrons and holes in the new structures, choice of interconnections, and design of circuits.

The advances in device performance that accompany miniaturization and integration have led to strong commercial competition that has given rise to increased awareness of the need to broaden the material basis. Often a significant improvement can be achieved by a bold choice of new materials. This is particularly the case when it comes to fulfilling the needs created by the attempts to integrate electronic and optical devices.

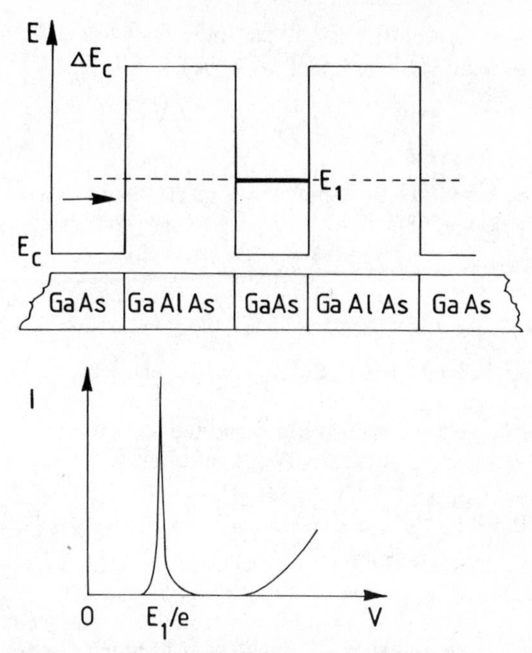

Fig. 11.11. (Top) A double barrier structure consisting of a thin GaAs layer sandwiched between two layers of GaAlAs. ΔE_c is the conduction band offset. E_c is the conduction band edge of GaAs. The barrier width is chosen so that electrons in GaAs cannot tunnel efficiently through the barrier unless their energy E coincides with E_1, which is the energy of the confined electron level in the middle GaAs layer. Since this GaAs layer is very thin (10–15 Å), the electron wave function peaks in the middle GaAs layer but its confinement is poor and it penetrates into the barrier layers. This ensures that the probability for electron tunnelling at E_1 (the so-called resonant tunnelling) is finite. (Bottom) A sketch of the current (I) versus voltage (V) curve expected for the double-barrier structure.

It means that, whereas in the pre-VLSI period the material basis was very narrow and almost exclusively dominated by silicon, with GaAs being used for limited specialized functions, in the post-VLSI period this picture will inevitably include a number of other materials and structures. For instance, substantial investment has been made into InP and InAs technology. These materials promise to provide superior mobilities and better control of hot-electron effects. They are therefore likely to replace GaAs in applications where fast switching is the main aim. The larger band offset may help in achieving smaller and more stable heterojunction structures and stretch the limits of VLSI towards the 0.1 μm mark.

Another class of developments concerns inter-chip connections, which will be revolutionized by advances in metal hardware (e.g. the high-temperature superconductor) and by new approaches to computer architecture (e.g. new forms of parallel processing).

Problems

11.1. Consider the heterojunction shown in Fig. 11.4 and estimate the energy of the lowest conduction miniband with respect to the bottom of the conduction band of GaAs (take $\epsilon = 13.1$, $T = 300\,\mathrm{K}$, and $x = 0.38$).

11.2. Compare the transit time of electrons in the active channel (of equal length) in the MOSFET and HEMT devices in Figs. 11.1 and 11.4.

11.3. Estimate the transmission coefficient for tunnelling at low temperatures of electrons into the base region of the hot-electron transistor shown in Fig. 11.6. The external electric field is $5 \times 10^5\,\mathrm{V\,cm^{-1}}$ and the alloy barrier contains 30% of aluminium. You can neglect the space charge contribution to the barrier potential and assume that the quasi Fermi energy of electrons lies 60 meV above the bottom of the conduction band of GaAs.

11.4. Estimate the base transit time of electrons in the hot-electron transistor shown in Fig. 11.6 and considered in Problem 11.3. Evaluate the shortening of the transit time on condition that the electron motion is entirely collision free and that the electron is accelerated by the applied field along the full length of its path.

11.5. Consider a periodic three-dimensional structure consisting of cubic quantum boxes of GaAs separated by GaAlAs barriers. If the separation between the boxes is the same as the length a of the box side, and if the concentration of donors is $10^{19}\,\mathrm{cm^{-3}}$, determine a so that there is on average one electron per box.

12
Application of semiconductor microstructures in optoelectronics

12.1. Heterojunction lasers

To understand how a heterojunction laser works, let us turn to Fig. 12.1. There we see on the left-hand side a simplified dispersion (band) diagram of a direct gap semiconductor (e.g. GaAs). Only the states near the fundamental gap E_g are shown. In an intrinsic crystal and at low temperatures, all valence states indicated by the solid horizontal lines are occupied by electrons, and all states in the conduction band above are empty. If we shine onto this crystal an intense beam of light whose energy is larger than the band gap energy, electrons will be excited across the gap, leaving behind empty slots (holes) at the top of the valence band. Following this excitation, the electrons in the conduction band relax to the bottom of the band by emitting phonons. This relaxation is very fast ($\simeq 10^{-12}$ s). The relaxation time for electron–hole recombination (across the gap) is of order $\sim 10^{-9}$ s, so that on this scale the time it takes to electrons to climb down to the bottom of the conduction band is negligibly small. Thus we can say that electrons are injected into the conduction band at a rate of I/eV per unit volume, where

$$\frac{I}{eV} = \frac{n}{\tau}, \tag{12.1}$$

n is the electron density, and V is the volume into which the electrons are confined after injection; τ is the relaxation time for electron–hole recombination and I is the electron current. At the steady state described in terms of these parameters, we have a reservoir of electrons in the conduction band and a corresponding reservoir of holes in the valence band.

We can define the 'effective' free-electron and hole Fermi levels E_{F_c}, E_{F_v} that will express the density of free electrons and holes, respectively, and which we can measure from the corresponding band edges. This is shown on the right-hand side of Fig. 12.1. Once this population regime is

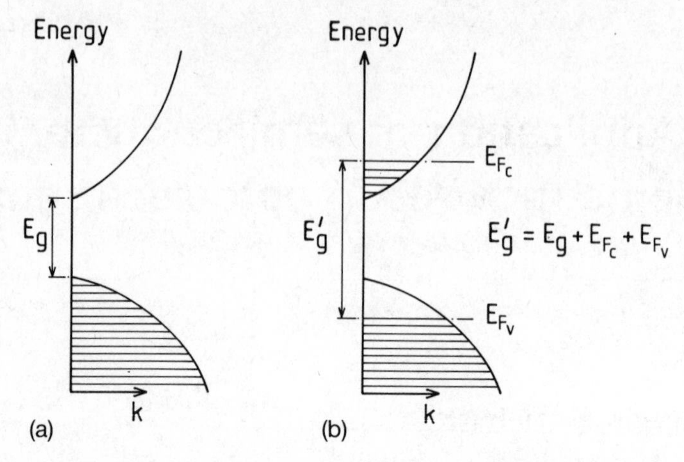

Fig. 12.1. (a) The band diagram of a direct gap semiconductor showing the top of the occupied valence band (horizontal lines) and the bottom of the empty conduction band, at $T = 0$ K. E_g is the forbidden gap. (b) Electrons have been transferred from the valence band into the conduction band (as indicated by the horizontal lines being removed to the bottom of the conduction band). E_{F_c} and E_{F_v} are the Fermi energies of the electron and hole reservoirs, measured from the corresponding band edges.

established, incident light of energy $\hbar\omega$ between E_g and $E_g + E_{F_c} + E_{F_v}$ indicated in the figure cannot be absorbed, because there are no electrons in the valence band and no suitable empty states in the conduction band for the process to occur. However, the electrons that have been transferred into the conduction band reservoir are free to emit a photon of energy $\hbar\omega$ and jump into one of the empty states in the valence band. The pumping beam of energy higher than the bandgap has created a situation in which the natural population of the semiconductor bands near the fundamental gap edges is 'inverted'. This situation is similar to the population inversion achieved in molecular lasers.

The diagram in Fig. 12.1 is highly idealized. At finite temperatures, there is a non-zero probability that some of the states near the top of the valence band are occupied. It is useful to define a minimum (threshold) value of the injection current, I_{th}, below which there is no significant emission of photons. In other words, if we want to achieve a meaningful photon flux generated by the recombination of the electrons with the holes from the inverted reservoirs, we must increase the power of the pumping beam so as to ensure that $I \geqslant I_{\text{th}}$. If, for example, the energy of the emitted radiation is $\hbar\omega_0 = E_g + E_c + E_v$, where $E_c = \hbar^2k^2/2m_e^*$, $E_v = \hbar^2k^2/2m_h^*$ are the energies of the initial and final states in the emission transition (k is zero at the band edge in both cases), then at

$I = I_{th}$ we must have the probability $f(E_c)$ that there is an electron at energy E_c larger than or equal to that at E_v.

By definition, $f(E_c) = f(E_v)$ at threshold, so that the light output power is proportional to $f(E_c) - f(E_v)$. The density of injected electrons can therefore be estimated theoretically from the explicit form of the Boltzmann distribution functions f, the density of states in the conduction and valence bands, and the response functions (susceptibility of Chapter 9) for propagation of electromagnetic waves in the system corresponding to the band diagram in Fig. 12.1. Near threshold, the relationship between output power (gain γ) and the injected electron density n is particularly simple and can be expressed by an empirical formula:

$$\gamma = A(n - n_{th}), \tag{12.2}$$

where n_{th} is a certain minimum (threshold) injected electron density. For GaAs lasers, n_{th} is typically 10^{18} cm^{-3} and $A \simeq 10^{-16}$ cm^2. It follows that we expect the gain γ to be in the range of several hundreds of cm^{-1}. This relationship is illustrated in Fig. 12.2, which gives empirical data for a GaAs laser taken at room temperature.

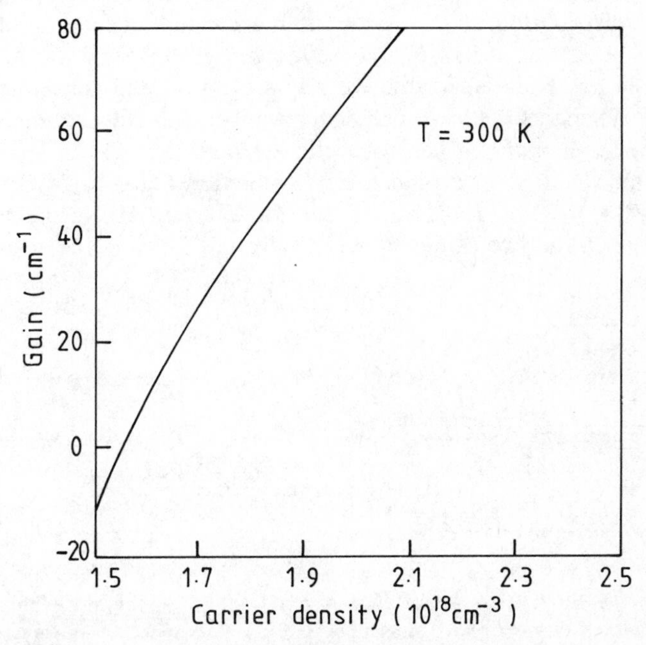

Fig. 12.2. A plot of the gain due to population inversion in a GaAs heterojunction laser as a function of the density of injected carriers.

This qualitative analysis points to the most important limiting factor for efficient generation of radiation from semiconductor lasers. In order to obtain large output power, we must increase the injection current, and we consequently require large n. As the figure shows, even at modest output power the carrier density needed is quite high. At such densities the material is heated and its reliability may deteriorate. The need to decrease the density of injection current is one of the main objectives in improving performance of semiconductor lasers.

The population inversion, on which the process in Fig. 12.1 depends, may be achieved at a p–n junction. In the heavy doping limit, the Fermi level of the n-type material forming the junction lies in the conduction band, whereas in the p-type material it is in the valence band. This is illustrated on the left-hand side of Fig. 12.3. When the junction is forward-biased, the electrons flow towards the p-type material and recombine in the junction region with holes. The energy released in this process is carried away as light (photons) as indicated on the right-hand side of Fig. 12.3.

The first semiconductor lasers were highly doped p–n junction structures described above. However, in this structure, electrons and holes are spatially separated from each other, and can only meet in the narrow region of the junction. Because of the high field and anisotropic gradient in charge distribution at the junction, the carriers diffuse rapidly away from the junction. This further separates electrons from holes and reduces the probability of radiative recombination and consequently the device performance. These problems are substantially diminished by employing semiconductor heterojunction structures.

The band diagram of a double-heterojunction GaAs laser is sketched in Fig. 12.4a. In Fig. 12.4b, we can see the change (Δn) in the refractive index between the two semiconductor materials forming the structure. A

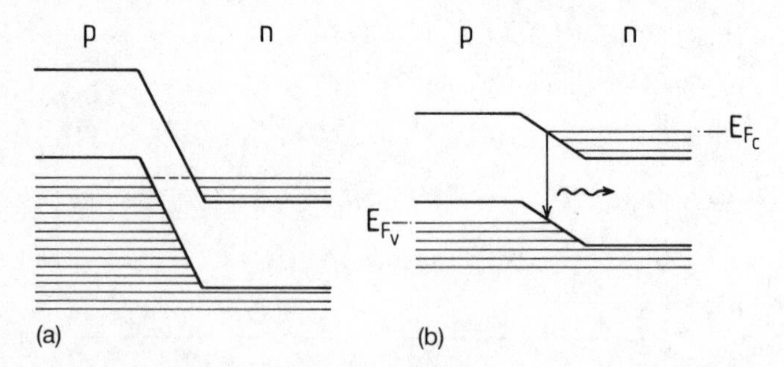

Fig. 12.3. (a) Band diagram of a p–n junction in the high-doping limit. (b) Junction with a large forward bias that is chosen so that the energy separation between E_{F_c} and E_{F_v} is larger than the band gap. Since electrons have higher mobility, we talk about emission of light due to electron injection.

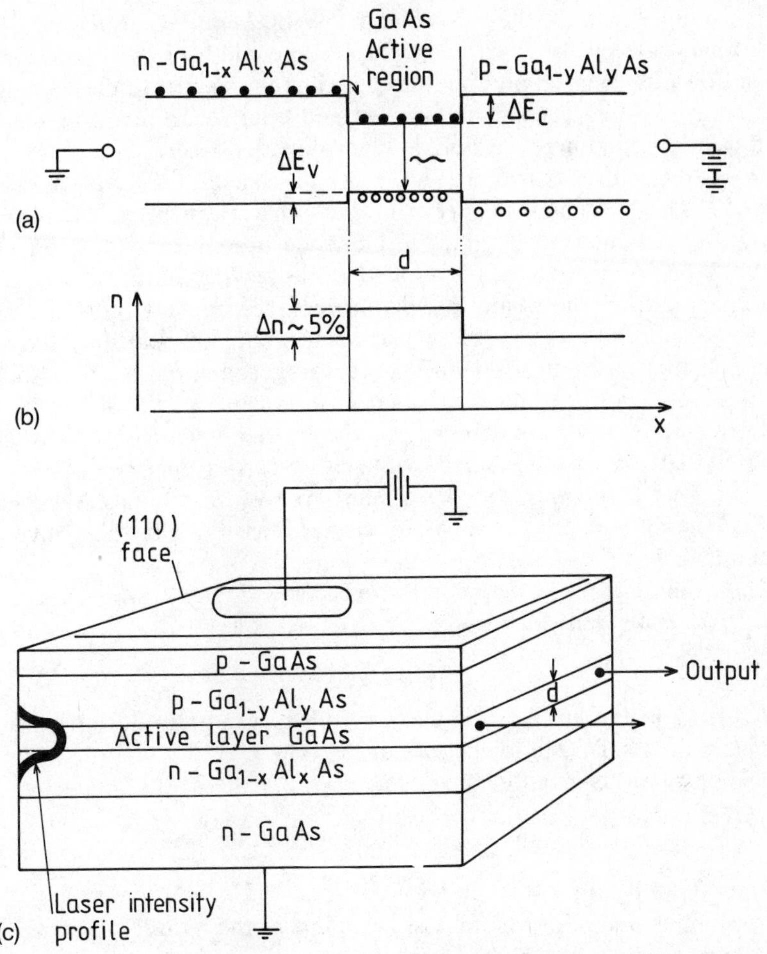

Fig. 12.4. (a) The band diagram of a double heterojunction laser. (b) The step-like variation of the refractive index n in the double-heterojunction laser structure. (c) The block diagram of a GaAs–$Ga_{1-x}Al_xAs$ laser.

schematic illustration of the physical layout of the device is given in Fig. 12.4c.

The active region is the GaAs layer. The radiation generated in this GaAs cavity has wavelength λ in the range $0.75\ \mu m < \lambda < 0.88\ \mu m$, depending on carrier density. In this structure (as in most practical applications), the GaAs layer is quite thick ($\simeq 0.1$–$1\ \mu m$) and the role of the two heterojunctions is simply to prevent electrons and holes from being spatially separated (i.e. there is no quantum-mechanical confinement leading to discrete levels). However, the structure does offer the opportunity to tune the wavelength of the emitted radiation by a judicious choice of the barrier height and the width of the middle layer. Quantum well lasers exploiting the quantum confinement effect discussed in Chapter 7 have been manufactured and operate on principles similar to those described here for double-heterojunction lasers.

In the structure shown in Fig. 12.4, the barrier material is heavily doped. The potential energy profile characteristic of a GaAs–$Ga_{1-x}Al_xAs$ doped heterojunction has been described before (e.g. Fig. 11.5). When a positive bias is applied to the device, electrons and holes are injected from the n- and p-type barriers, respectively, into the active GaAs layer. Once there, these carriers cannot escape, since they are confined by the band offset and space charge potential barrier (in Fig. 12.4, the effect of band bending is neglected for the sake of simplicity and only the band offset contribution to the barrier height is shown). This ensures that for a fixed external voltage the carrier density in the active layer is only a function ($\sim 1/d$) of the layer width d. Since the output power (gain) γ is proportional to carrier density, γ is also inversely proportional to d.

The difference between the refractive indices of GaAs and $Ga_{1-x}Al_xAs$ depends on x as

$$n(\mathrm{GaAs}) - n(\mathrm{Ga_{1-x}Al_xAs}) \simeq 0.62x, \tag{12.3}$$

and it is large enough that the GaAs also acts as a natural waveguide for the generated light (see lower part of Fig. 12.4).

The net gain achieved in the device is

$$\frac{g}{g_0} = \frac{\text{power generated per unit length}}{\text{power carried by the beam.}} \tag{12.4}$$

Let us define α_n and α_p as the loss constants of the n- and p-type regions. The power carried by the beam is

$$g_0 = \int_{-\infty}^{\infty} |E|^2 \, dz, \tag{12.5}$$

where z is perpendicular to the interface plane and $|E|^2$ is the field intensity. Hence,

$$\frac{g}{g_0} = \gamma \int_{-d/2}^{d/2} |E|^2 \, dz - \alpha_n \int_{d/2}^{\infty} |E|^2 \, dz - \alpha_p \int_{-\infty}^{-d/2} |E|^2 \, dz - \frac{\ln(R)}{L}, \tag{12.6}$$

where γ is the gain due to population inversion and d is the width of he active GaAs layer shown in Fig. 12.4. A rough graph of intensity $|E|^2$ across the GaAs layer (perpendicular to the interfaces) is shown in Fig. 12.5. We can see from eqn (12.6) and Fig. 12.5 that although the carrier

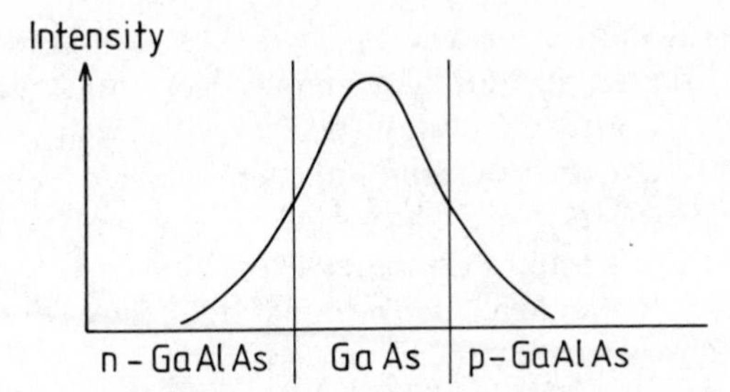

Fig. 12.5. The profile of the electromagnetic field generated in a double-heterojunction laser and propagating as indicated in Fig. 12.4c (parallel to the GaAs–$Ga_{1-x}Al_xAs$ interfaces).

density, and consequently γ, increases as $1/d$, with decreasing d a larger fraction of the generated beam is carried in the alloy and not in GaAs. For small d, the $1/d$ dependence is lost because of poor confinement of the light beam.

At threshold, we have

$$\gamma_{\mathrm{th}} \int_{-d/2}^{d/2} |E|^2 \, \mathrm{d}z = (\alpha_{\mathrm{n}} + \alpha_{\mathrm{p}}) \int_{d/2}^{\infty} |E|^2 \, \mathrm{d}z + \frac{\ln(R)}{L}. \tag{12.7}$$

The last term on the right was added to account for losses incurred by the light beam in reflections. R is the reflectivity and L is the length of the beam path in the waveguide. This term is obtained as follows. The loss due to reflections can be written in the form $R\exp(-2L\alpha)$. The factor 2 accounts for the fact that the beam passes through the medium twice. The gain along the same path is $R\exp(2L\beta)$. Since the product of these two expressions must equal 1, we obtain for the net loss

$$\alpha - \beta = -\frac{\ln(R)}{L}. \tag{12.8}$$

We have assumed that the reflectivities R_1 and R_2 of the two mirrors at the ends of the length L are the same, i.e. $R_1R_2 = R^2$, where

$$R = \left(\frac{n-1}{n+1}\right)^2, \tag{12.9}$$

and n is the refractive index.

We can now estimate γ_{th}. Let us take for the refractive index of GaAs $n = 3.6$ and $L \simeq 500\ \mu\mathrm{m}$, so that $\ln(R)/L = -23\ \mathrm{cm}^{-1}$. For large $d(\simeq 1\ \mu\mathrm{m})$, the leakage of intensity out of the GaAs waveguide is small and the integral on the left-hand side of eqn (12.7) can be set to 1. The integrals associated with α_{n} and α_{p} on the right of eqn (12.7) are therefore small. The threshold value of gain γ_{th} is then about $23\ \mathrm{cm}^{-1}$. In practice, the losses due to leakage that we neglected here, and also an additional loss due to scattering at imperfections at interfaces, add another $10\,\mathrm{cm}^{-1}$. If we take the more realistic value of $\gamma_{\mathrm{th}} \simeq 33\,\mathrm{cm}^{-1}$, we find from Fig. 12.2 that threshold injection carrier density is about $1.7 \times 10^{18}\,\mathrm{cm}^{-3}$.

Under steady-state pumping conditions, the rate at which carriers are injected into the active layer must equal the rate of electron–hole recombination. The injected current density is then

$$j = en\frac{d}{\tau}. \tag{12.10}$$

Using the above value of carrier density n and $\tau \simeq 5 \times 10^{-9}$ s, we obtain at threshold for $d = 1\ \mu\mathrm{m}$,

$$j_{\mathrm{th}} \simeq 5 \times 10^{3} \quad \mathrm{A\,cm}^{-2}. \tag{12.11}$$

The double-heterojunction structure in Fig. 12.4c lacks the means for confining the optical beam and current in the lateral y direction. This means that the mode propagating in the GaAs waveguide is not very stable and the width of the laser line is unacceptably high. To overcome this problem, other structures have been employed in which the GaAs active layer is 'buried' (surrounded from both x and y directions by the low-refractive-index $\mathrm{Ga}_{1-x}\mathrm{Al}_x\mathrm{As}$ alloy barrier material). To fabricate such a structure, the three layers, n-$\mathrm{Ga}_{1-x}\mathrm{Al}_x\mathrm{As}$, GaAs, and p-$\mathrm{Ga}_{1-y}\mathrm{Al}_y\mathrm{As}$ are grown, as before, on a high-quality GaAs substrate by one of the epitaxial growth techniques described in Chapter 10. The structure is then etched through a mask down to the surface of the substrate, leaving stand a rectangular mesa of the three layers. This mesa is then regrown and a 'burying' $\mathrm{Ga}_{1-x}\mathrm{Al}_x\mathrm{As}$ material is deposited so that, as a result, the active GaAs layer is surrounded by the barrier on two sides. In these structures, the threshold current is reduced to less than 1 mA.

In the heterojunction structure of Fig. 12.4c, the stimulated coherent radiation generated in the GaAs layer is achieved by turning the sandwich semiconductor system into a resonator. The photons generated by recombination across the forbidden gap of GaAs are returned by the cavity only to stimulate electrons in the conduction band to emit photons of the same energy and phase. The laser described in Fig. 12.4c can be viewed as an oscillator powered by the pumping source that maintains the population inversion. The multiplicity of the reflections in the laser cavity is reminiscent of the well known Fabry–Perot resonator. Although the signal amplification from a single pass of the beam through the medium is small, after a number of such passes the net gain is considerable. A steady-state as described above is obtained when the losses incurred by the beam equal the optical gain.

If there are no mirrors, the radiation is simply that produced by electrons that emit at random photons of the band gap energy, and the system is called a light-emitting diode (LED). This spontaneous emission is again approximately monochromatic but is incoherent, since the photon feedback that acts to select the coherent wave through the avalanche re-emission is missing.

Since there are no reflections, the injection current density required to operate an LED is lower. The need to maintain high current in the material and the high level of energy dissipation (i.e., heating of the lattice) also leads to shorter lifetime of semiconductor lasers and lower reliability compared to LEDs. In fact, the LED structure is simpler, and consequently cheaper to make.

The obvious advantage of the laser is the possibility of coherent transmission and detection. The transmitted information can be coded as phase, amplitude, and frequency changes of the coherent light waves. However, to take full advantage of coherent transmission, it is necessary to reduce the laser line width, which in double-heterostructures described above may be as large as 100 MHz. Such large spectral widths arise mainly from fluctuation in the refractive index of the material and from the strong background spontaneous emission. The linewidth can be reduced to about 10 kHz by adding an external cavity to the laser and by careful control of material and manufacture quality, temperature fluctuations, and the non-linear effects in the cavity.

The heat dissipation is a source of electron density fluctuation because energy is exchanged between the electron reservoir and the lattice. Figure 12.6 shows the deterioration of the laser linewidth as a function of increasing power output and temperature.

Another source of problems is the poor definition of the polarization (the mode of propagation) of the beam. Optical repeaters have been equipped with a multitude of optical devices designed to control and correct the mismatch between the transmitter and receiver. A block diagram of an optical repeater is shown in Fig. 12.7.

The first heterojunction lasers and LEDs were made of GaAs and GaAlAs alloys. However, the wavelength of the radiation generated in these structures is in the range $\simeq 0.8\,\mu$m and is determined by the band gap of GaAs. Although it is possible to alter this wavelength by employing quantum wells and superlattices, the range of wavelengths covered is still quite small. For many applications, monochromatic sources of coherent or incoherent radiation of different wavelengths are required. It follows that other materials must be sought to satisfy such needs.

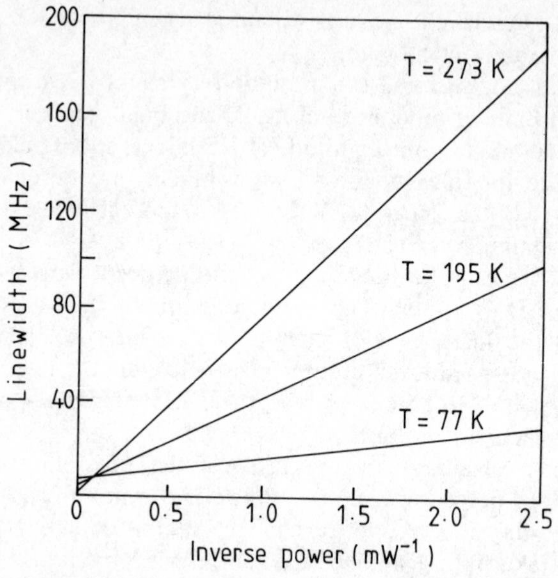

Fig. 12.6. The line width of a GaAs-$Ga_{1-x}Al_xAs$ laser as a function of inverse power generated by the laser and of temperature T.

In Fig. 12.8, we can see a summary of the most important materials that have been used to grow heterostructures such as that in Fig. 12.4c, and the wavelengths that can be covered by these structures. The substrate availability is an important technological consideration and the figure shows that a variety of substrate materials must be available for the band structure engineering effort to succeed in covering the full wavelength range. The most significant bands are the 1.55 μm range, which is

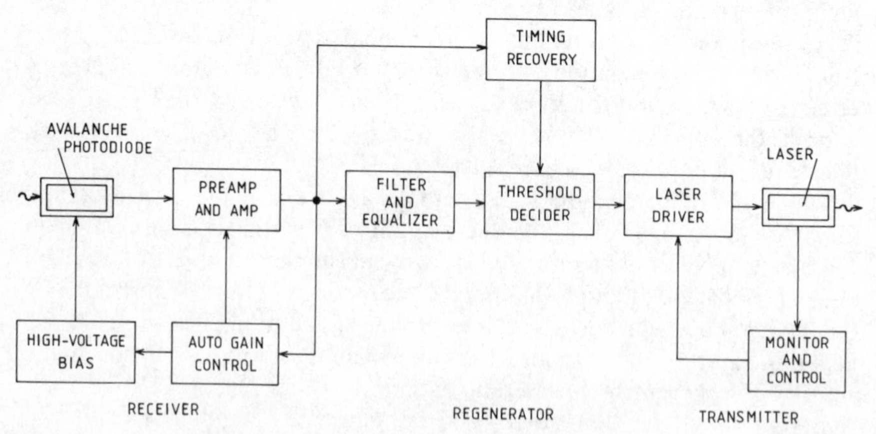

Fig. 12.7. A block diagram of a light repeater.

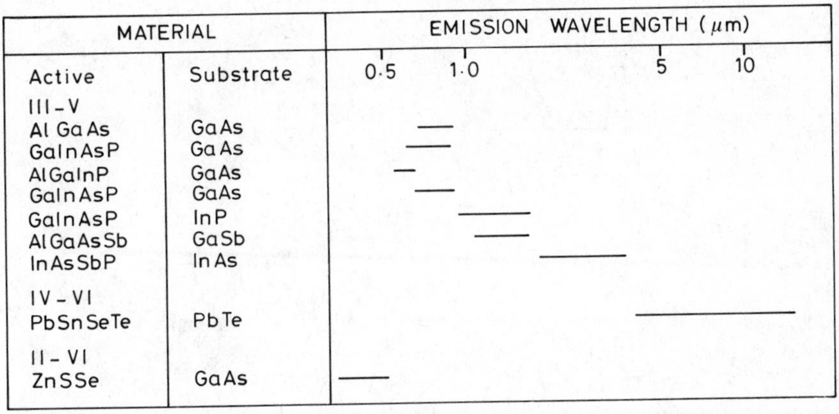

MATERIAL		EMISSION WAVELENGTH (μm)
Active	Substrate	0.5 1.0 5 10
III–V		
AlGaAs	GaAs	
GaInAsP	GaAs	
AlGaInP	GaAs	
GaInAsP	GaAs	
GaInAsP	InP	
AlGaAsSb	GaSb	
InAsSbP	InAs	
IV–VI		
PbSnSeTe	PbTe	
II–VI		
ZnSSe	GaAs	

Fig. 12.8. Semiconductor materials for light sources, showing the range of light wavelengths that can be obtained with these materials.

ideal for optical fibres, the far infrared band for detector applications, and the visible light band for displays. For instance, while the InP technology is best suited for the 1.55 μm range, the infrared and the visible bands might be best covered with narrow (HgTe–CdTe) and wide (ZnSe–ZnS) gap II–IV semiconductor microstructures, respectively. Thus the needs of optoelectronics greatly broaden the materials basis of semiconductor technology.

12.2. Photodetectors

Processing of optical signals requires photodetectors that convert optical signals into electrical currents. This is because amplification and data processing is done by electronic devices. The performance of photodiodes is measured by the speed of response, sensitivity, low noise, and high gain.

There are two types of photodiodes. The first devices were just p–n junction diodes in which the incoming photons are absorbed by electrons in the valence band. The current of the electron–hole pairs created in this process is then amplified and used to drive electronic data processors. The detection process is simply the complemetary process to the generation of light described in Section 12.1. For every incoming photon, the diode can at best produce one electron–hole pair. These devices were made of germanium or silicon.

More efficient devices can be made by exploiting the hot-electron avalanche multiplication that occurs in biased heterojunction ('staircase') structures. The functioning of such structures was described in Chapter 8.

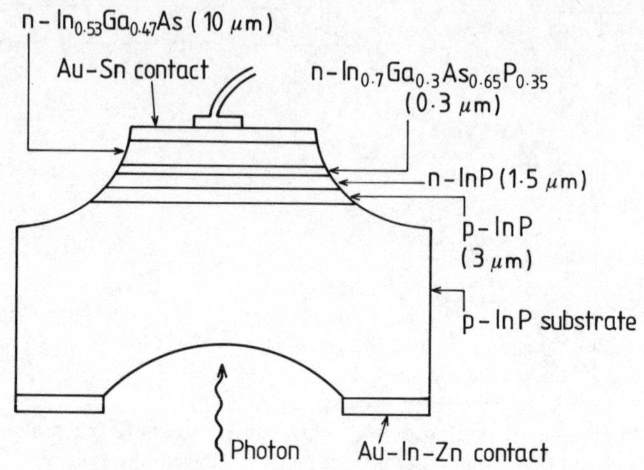

Fig. 12.9. Block diagram of an avalanche photodetector.

The choice of heterostructure may enable us to select the best wavelength. An example of an avalanche heterojunction detector is shown in Fig. 12.9. In this case, a relatively thick layer of $In_{0.53}Ga_{0.47}As$, lattice-matched to the InP substrate, is used to absorb the incoming light. The holes generated by the absorption are driven to the high-field region near the p–n junction, where they are multiplied. The substrate is transparent to photons absorbed in the active layer and so are the epitaxial n- and p-doped InP layers in which the multiplication occurs.

12.3. Optical fibres

The invention of the laser in the early 1960s underlined the need for new communication lines. The result of a concerted effort of two decades is a glass optical fibre waveguide that has become a backbone of modern information technology.

A rough sketch of such a waveguide is shown in Fig. 12.10. The core of glass fibre is made of high-quality silica (SiO_2) of refractive index n_1. It is surrounded by a cladding material of somewhat lower refractive index (n_2). A light beam impinging on the interface between the cladding and the core at a sufficiently small angle is totally reflected towards the centre of the core and the light is effectively trapped and guided along the fibre axis. A light source such as a semiconductor laser feeds modulated light into the fibre. It is essential that the fibre is capable of carrying light over a distance of many kilometers without it suffering strong attenuation, so that the signal can easily be detected at the endpoint by a photodiode.

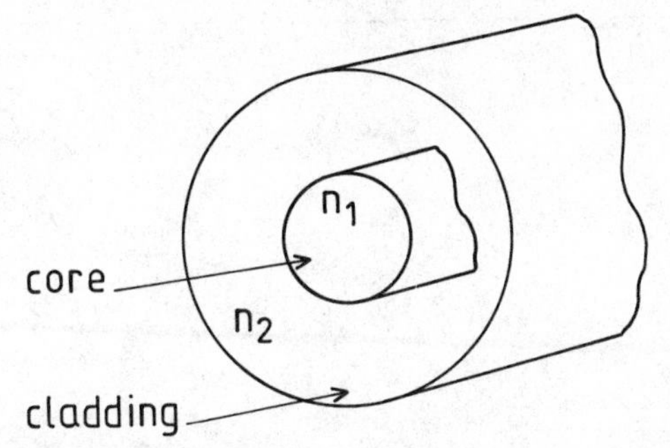

Fig. 12.10. A sketch of an optical fibre consisting of a glass core of refractive index n_1 and cladding of a slightly lower index n_2.

This diode converts the signal into an electrical current. The electronic equipment in a repeater (Fig. 12.7) amplifies the signal, which is then converted by a laser into a light beam and fed into the next waveguide. For the line to work efficiently, the properties of the fibre and of the optical devices (lasers or LEDs) operating the line must be well matched.

The degree of attenuation suffered by the light beam in the fibre depends on the microscopic physical properties of the silica used to make the fibre. The empirical loss-versus-wavelength curve for a high-quality fibre is shown in Fig. 12.11. The loss is given in decibels per kilometre and also (right ordinate) in 1/km, implying that the light intensity decreases exponentially with distance. The solid line represents the usual commercial material. The dotted line is the best result achieved. The solid line has a principal minimum at about 1.55 μm, and secondary minima at shorter wavelengths. The peaks at ~0.95 and 1.38 μm correspond to the (stretching) vibrational frequencies of OH groups that are common contaminants in silica. The peak at 1.25 μm is related to the vibrational frequency of a SiO_2 molecular tetrahedron. The most advanced techniques of silica making can, in fact, reduce the absorption due to these vibrations so much that the loss curve approaches the fundamental limit. In the short-wavelength range, the limiting mechanism is light scattering (the so-called Rayleigh scattering, which is inversely proportional to the fourth power of wavelength). In the long-wavelength limit, the attenuation rises because of the Si–O–Si stretching vibrations. In any case, the lowest loss occurs at around 1.55 μm. The remaining difference between the dotted line and the fundamental limits (broken lines) is due to small fluctuations in the refractive index associated with inhomogeneities in the silica that were introduced during growth.

A light beam propagating in a dielectric suffers dispersion that alters the original shape of the beam. This effect also depends on the wavelength of the transmitted wave and must be taken into account. One

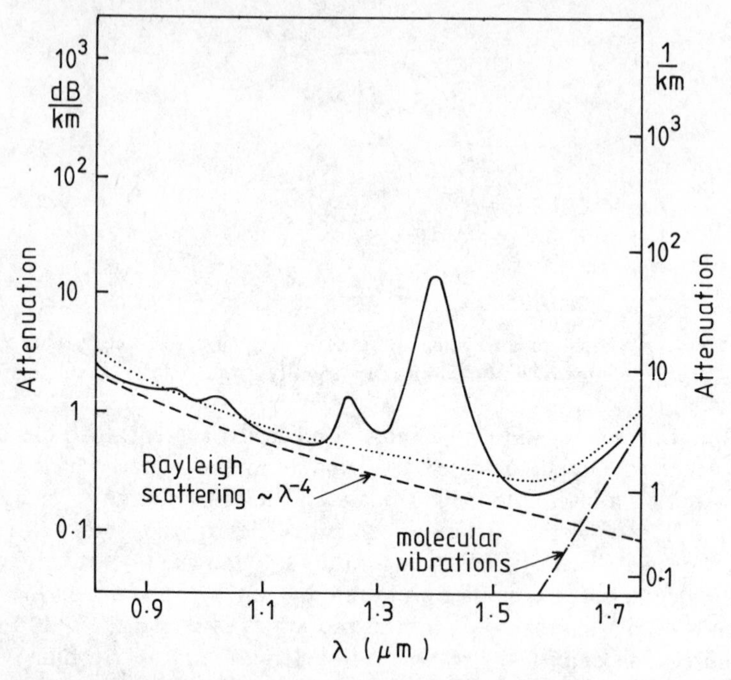

Fig. 12.11. Attenuation in an optical fibre as a function of the wavelength of the propagating beam. The dotted line is the state-of-the-art result. The solid line represents a typical high-quality fibre. The broken lines show the fundamental limit for silica fibres.

important source of dispersion is so-called modal dispersion. Several wave modes can propagate in fibres with large core diameter. This can be understood simply if we recall that the propagation of light in the core has the form of a wavefront, as indicated in Fig. 12.12. For the wavefront to be preserved at the end of the optical path over which it is supposed to travel, the wavefront must reconstruct itself after reflections. These reflections are illustrated in the figure at points A and B. In particular, the total phase difference between A and B must be a multiple of 2π. This means that, if m is an integer, we can write

$$-2\pi m = \mathrm{AO} - \mathrm{AB} + 2\delta, \tag{12.12}$$

where AO − AB is the path difference shown in the figure and δ is the change of phase associated with the reflection at the interface between the core and the cladding. This defines the number of modes that can propagate in a fibre. Each mode is characterized by a profile of the electric field in the direction perpendicular to the fibre axis, as shown on the right-hand side of Fig. 12.12.

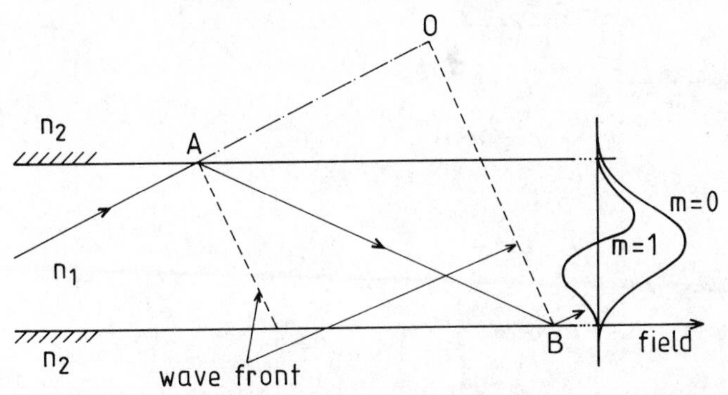

Fig. 12.12. The wave reflections in an optical fibre waveguide such as shown in Fig. 12.10. On the far right is shown the field profile for the first and second modes ($m = 0$ and $m = 1$) of propagating waves.

The phase change δ depends on the quality of the cladding interface. In particular, if the interface allows the beam to penetrate into the cladding material, the optical path differs from the ideal one shown in Fig. 12.12. If several modes propagate in the fibre (i.e. if the core geometry is such that $m > 0$), the fibre is called a multimode fibre. The final beam is a superposition of waves of different modes and its overall shape is altered as a result. The number of modes can be reduced by reducing the core diameter until at about 5–10 μm only the totally symmetric mode $m = 0$ remains. A typical diameter of the cladding is 120 μm.

In a single-mode fibre there is also some dispersion, which is determined by the geometry of the core and by the refractive indices n_1 and n_2 (the 'waveguide' dispersion). Furthermore, in a dielectric, the group velocity and the refractive index are in general functions of wavelength (Chapter 9) that depend on the microscopic properties of the dielectric (the 'material' dispersion). It is important to distinguish between these contributions in the assessment of the fibre quality, since their wavelength dependences differ. The empirical results showing the magnitude of the dispersion, which is usually measured in picoseconds per kilometre-nanometre, are presented in Fig. 12.13. This curve indicates that there is practically no dispersion at wavelengths around 1.55 μm.

The fabrication of the glass fibre is done by depositing the glass from vapour phase at high temperatures. The step in the refractive index of the silica needed to provide the waveguide effects is achieved by adding GeCl to the vapour. Fibres of 20–100 km length are produced by pulling the glass from the furnace and coating it with a polymer for external protection. The fluctuation of the core diameter and the roundness of the

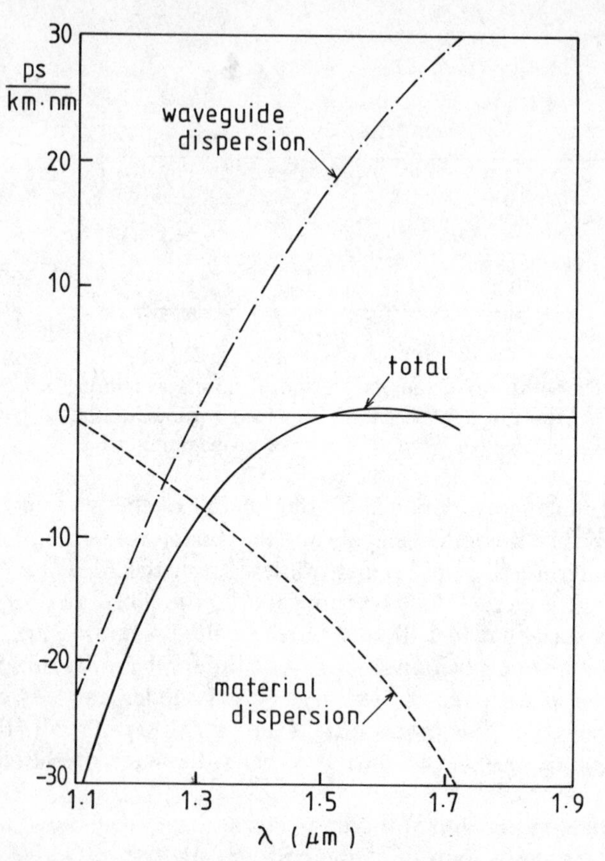

Fig. 12.13. The material and waveguide contributions to the dispersion of a propagating beam in an optical fibre. The total dispersion has a minimum at wavelengths around 1.55 μm. The dispersion is represented as $(1/L)\, d\tau/d\lambda$, where τ is pulse delay, L is fibre length (in km) and λ is wavelength of the propagating beam (in nm).

core and cladding are fabricated with the precision of 1 μm per several kilometres. The refractive index of the core is, in fact, designed to have a graded cross-sectional (radial) profile so that the higher modes, which travel at larger angles to the fibre axis and have the longest optical path, are delayed. This grading is achieved by adding GeO_2. The refractive index varies so that it parabolically decreases towards the interface with the cladding. The best fibres operated at around 1.55 μm have losses reduced to 0.5–0.2 dB km^{-1} (Fig. 12.11).

At the long-wavelength limit, there is the possibility of reducing the losses even further. For example, the fibre could be made of halides such as KCl or CsI. The losses expected in these materials are of the order of 10^{-4} dB km^{-1} at $\lambda \simeq 5$ μm. This is because, as we noted in Chapter 4, the vibrational frequency is inversely proportional to the mass of the vibrating atoms. These materials consists of heavier atoms, so that the characteristic frequency shifts to longer wavelength. Also, the Rayleigh scattering is weak at large λ. These materials, unlike silica, are single crystals, so that the material purity is better controlled. However, the cost of developing a new fibre technology is prohibitive.

12.4. Optical switches

Photons are massless and chargeless particles and, unlike electrons, do not interact with each other significantly. As a result, it is possible to send intense beams of light through an optical fibre over distances of many kilometres without distortion of the signal. Another way of taking advantage of this property of photons is to use optical beams as carriers of information in parallel processing. Many different beams can be used in the same space without any danger of unwanted photon–photon collisions. On the other hand, these are precisely the reasons why photons are unsuitable for amplification and switching.

In Chapter 9 we related switching to bistability. We expressed the existence of bistability in terms of a finite 'non-linear' contribution to the refractive index. This is the change in the refractive index that depends on the intensity of the applied optical beam. In most materials found in nature, this field-dependent component of the refractive index is very weak, and when we consider propagation of light waves, we normally ignore its existence. However, in substances with large polarizability, and at applied fields of high intensity, this non-linearity can be used to make a bistable device (an optical switch). As we showed in Chapter 9, it can also be used to make modulators, frequency converters, and mixers, and other 'non-linear' optical devices.

The most popular material used in commercially available optical devices is lithium niobate ($LiNbO_3$). A thin film of titanium is deposited to make a waveguide pattern on the crystal. The refractive index of titanium is higher and light is confined in the crystal. Two such waveguides can be positioned close to each other on the crystal surface, as indicated in Fig. 12.14. An external field is applied across the waveguides. The induced change in the refractive index causes the light beam to switch from one guide to the other. The 'on' and 'off' positions are on the right- and left-hand sides of this figure, resp.

The excitonic absorption in GaAs quantum well structures discussed in detail in Chapter 9 can also be used to make an optical switch. The

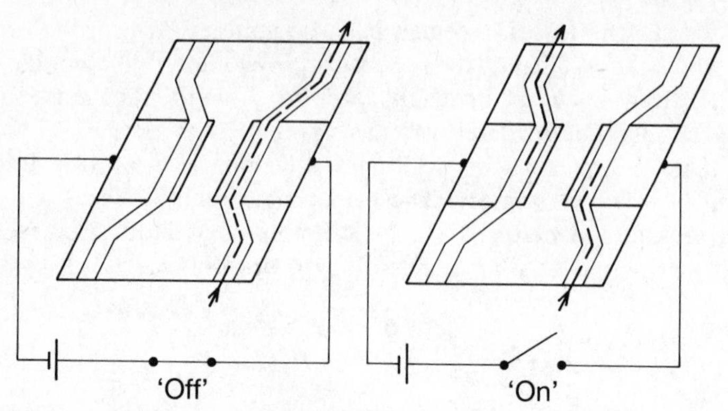

Fig. 12.14. A simple optical switch that can redirect an optical beam from one waveguide to another by the application of an external voltage. 'On' and 'off' positions are shown.

non-linearity in the optical response is built into this system, in that the strength of the absorption near the band edge depends on the existence of localized excitonic levels lying well below the band edge. Therefore, when these excitonic states are destroyed or their depth is reduced significantly by an external electric field (see, for example, Fig. 9.6b), the response (absorption) is greatly reduced. In the example of non-linear (switching) behaviour given in Chapter 9, we used two light beams (a holding and a switching beam) to turn the systems from a low-transmission position to a high one (Fig. 9.5).

A GaAs quantum well heterostructure can also be operated as a self-electrooptic gate. A schematic representation of the device is shown in Fig. 12.15. We begin by applying a voltage just large enough that the excitonic levels are shifted to a shallow position and the excitonic

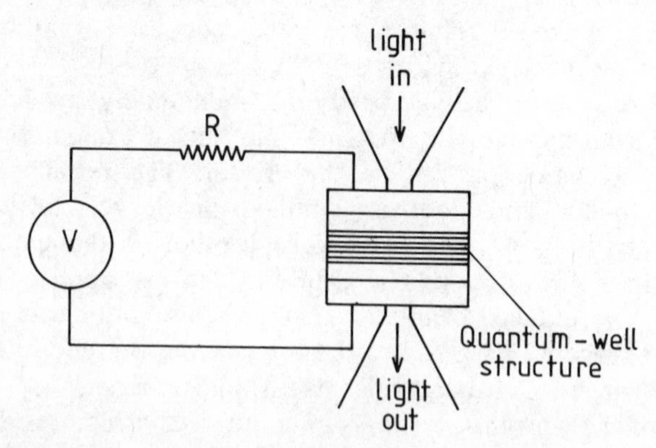

Fig. 12.15. A quantum well switch described in the text.

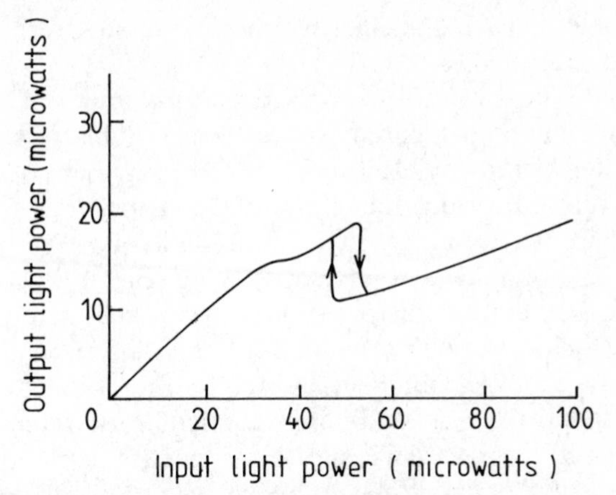

Fig. 12.16. The input–output power curve for the quantum well switch shown in Fig. 12.14, exhibiting hysteresis characteristic of bistable (non-linear) optical materials.

absorption is very low. Then we switch a light beam onto this device. The relationship between the input and output power of the device is illustrated in Fig. 12.16. As we increase the beam intensity, the light generates a photocurrent in the sample. This eventually reduces the voltage across the device. As a result of this reduction in the electric field, the excitonic levels become deeper and absorption in the sample increases. However, if we continue to increase the intensity, the increase in absorption leads to an increase in the photocurrent and this in turn reduces the voltage and the absorption. We can now start decreasing the input intensity. The return to the state of high transmission does not occur at the same intensity because the device is now absorbing. Consequently, the output–input curve exihibits the hysteresis characteristic of non-linear optical materials (bistable elements).

12.5. Applications of photonics in information technology

The field of photonics, a label often used to embrace a multitude of physical principles and devices involving (linear and non-linear) optical phenomena, has grown enormously since the 1960s when the laser and non-linear optical devices were first demonstrated. The rate of growth and technical progress has accelerated even further with the advent of semiconductor heterojunction microstructures. The revolutionary role of these microstructures in photonics is analogous to that in electronics. As we have seen in this course, practically all classes of optical devices are affected by the emergence of new semiconductor structures. These new devices are potentially more efficient, more compact, and better suited for mass production.

However, it is important to recognize that photonics is in an infantile state of development compared to electronics. This is so even if we ignore the long history of electronic circuitry making and the mass of knowledge acquired during the pre-semiconductor era. To understand the relationship between electronics and photonics, we only have to compare the age and state of development of silicon technology, which nourishes the success of microelecronics, with the maturity of GaAs technology, which gave birth to modern photonics.

The first stage of development of photonics has been a process of perfection of optical fibre and laser communication technology, and integration of these optical devices with the electronic processing circuits. The fulfilment of this stage may coincide with the achievement of the physical limit of silicon VLSI technology in particular, and the era of the conventional transistor circuitry in general. Even in this restricted view of photonics (i.e. as providing communication lines), substantial progress must be made before the full potential of optical fibres and semiconductor lasers is realized. Silica optical fibre manufacture is approaching the fundamental limit of perfection. However, lasers must be improved at least in two ways.

Firstly, they must be able to generate light of much narrower line width. Although in principle a semiconductor laser should produce monochromatic coherent light, in practice the light generated spreads over a wide band of frequencies. Light waves of different frequency travel in an optical fibre at different speed and the repeater consequently receives a distorted signal. As for coherence, we have hardly begun to develop and exploit the coherent nature of laser light. In fact, in most practical applications of optical fibres, incoherent LED sources are used because they are cheaper and more reliable.

Secondly, it is useful to have a tunable source. Such lasers would make a better use of the fibres' capacity because many lasers operating at different frequencies can simultaneously transmit independent signals over the same fibre. This is called frequency multiplexing. The beam of a tunable laser can also carry more information, since the signal can be frequency modulated.

An improvement in linewidth and frequency tunability may then be followed by an increase in the pulse capacity. The limit of the pulse capacity achievable with a semiconductor laser depends on the maximum power the laser can generate, the number of the photons necessary to define a pulse, and on the level of attenuation and dispersion in the fibre. As a measure of performance, the factor of interest is the product of the pulse rate and the maximum distance between repeaters, i.e. the distance the light signal can travel without amplification. The limit expressed in pulse-kilometres is about 10^{15}, which is about three orders of magnitude above the best laboratory results reported in 1986. The repeaters can at best be positioned about 60 km apart, and about 100 photons are needed to define a pulse.

Because a laser can only pulse as fast as the electronic circuits used to process the signal permit, substantial improvement in the speed of electronic circuits is also necessary. The fastest silicon circuits are much too slow for this task. The processing photonic devices (i.e. optical switches) described in this chapter are still too large, too slow, and too power-hungry to compete with transistors, and exist only as laboratory prototypes.

It follows that the second stage of development of photonics, during which we might expect the emergence of photonics circuitry independent of electronics, seems unthinkable without a concerted effort to develop an optical device equivalent to the electronic transistor—an efficient and small switch that does not dissipate a prohibitive amount of energy. We have seen that even the optical switches based on heterojunction microstructures require a continuous supply of power (e.g. a holding beam). Because the optical non-linearity is so small, very large intensity of the applied field is needed. Such devices are unsuitable for making high-density circuitry such as would be required for an optical computer. Similar requirements apply to other photonic devices required to amplify and modulate light beams or alter their frequency. It would appear that new physical concepts of optical switching must be developed to make optical computing a realistic proposition.

Problems

12.1. Estimate the separation of light modes propagating in a GaAs heterojunction laser shown in Fig. 12.4. Assume that the laser is emitting light of wavelength $\lambda = 0.9\ \mu m$ and that the separation between the sides forming the cavity is $L = 500\ \mu m$.

12.2. Show that the gain due to population inversion in a GaAs heterojunction laser is proportional to $(\omega - E_g/\hbar)^{1/2}$, where E_g is the forbidden gap of GaAs.

12.3. Estimate the change in the threshold current if the active layer of the device considered in Problem 12.1 is made of $Ga_{0.62}Al_{0.38}As$.

12.4. The onset of the vibrational lattice modes is a limiting factor determining attenuation in optical fibres at long wavelengths.
(a) Make a rough estimate of the ratio of the characteristic vibrational frequencies in SiO_2 and KCl, assuming that the difference between the elastic constants of these two materials can be ignored.
(b) Estimate the corresponding change in the wavelength of the vibrational mode.

12.5. Show in qualitative terms that quantum wire and dot structures might be used to manufacture optical switches that would operate at higher temperatures than the quantum well structures described in the text.

12.6. Estimate the order of magnitude of the ratio $\chi^{(3)}/\chi^{(1)}$ for a typical bulk semiconductor due to bound valence electrons. State what units are used. Express n_2 of eqn (9.40) in SI and CGS units (the so-called electrostatic units that are often used in scientific texts on optical properties; see table of fundamental constants).

Appendix 1

Hydrogenic levels

Practically all the quantum mechanics required in this course can be reduced to two most elementary problems. One is the solution of the Schrödinger equation for a free hydrogen atom. The other, outlined in Appendix 2, is that of an electron in a one-dimensional step-like potential (i.e. the "particle-in-a-box" problem).

To describe the motion of an electron in the field of an attractive Coulomb potential $-e/(4\pi\epsilon_0\epsilon r)$ associated with the hydrogen nucleus (proton), we must evaluate both the electron kinetic and potential energies. The classical energy is $p^2/(2m) - e^2/(4\pi\epsilon_0\epsilon r)$, where $p = mv$ is the linear momentum and r is the distance between the electron and proton. However, only the energies associated with the orbital radius a such that the angular momentum satisfies the condition $mva = nh/(2\pi)$, where $n = 1, 2, 3, \ldots$ and h is Planck's constant, fit the observed hydrogen optical spectrum. Thus the allowed energy levels E can be labelled by the integer n, which is called the principal quantum number. They form a discrete set

$$E_n = -\left(\frac{e^2 Z}{4\pi\epsilon_0\epsilon}\right)^2 \frac{m}{2\hbar^2 n^2} = -\frac{\hbar^2}{2ma^2}, \tag{A1.1}$$

where $Z = 1$ and $\epsilon = 1$ in the case of hydrogen in free space, and $\hbar = h/(2\pi)$.

Since electrons behave as both particles of mass m and as waves (e.g. we can observe electron diffraction), we must define the electron wave function ψ such that $|\psi(x, y, z)|^2$ is the probability of finding an electron at a point with Cartesian coordinates x, y, and z. In order to find ψ, we must solve the Schrödinger equation, which is a second-order partial differential equation of the form

$$\left\{\frac{\hat{p}^2}{2m} - \frac{Ze^2}{4\pi\epsilon_0\epsilon r}\right\}\psi = E_n\psi, \tag{A1.2}$$

where $\hat{p}$ is now an 'operator' defined as $p_x = -i\hbar(\partial/\partial x)$, $p_y = -i\hbar(\partial/\partial y)$

and $p_z = -i\hbar(\partial/\partial z)$. $r = (x^2 + y^2 + z^2)^{1/2}$ is the radial coordinate defining the electron position relative to the nucleus at $x = y = z = 0$. Since the squared modulus of the wave function is a probability, we must require that

$$\int_{-\infty}^{\infty} |\psi(x, y, z)|^2 \,\mathrm{d}x\, \mathrm{d}y\, \mathrm{d}z = 1. \tag{A1.3}$$

We say that the wave function is normalized to 1. We expect ψ and its first derivative to be continuous everywhere.

Suppose that we know the functional form of ψ. We can obtain (the expectation value of) the electron energy by substituting for ψ into eqn (A1.2), multiplying this equation from the left by the complex conjugate of the wave function (i.e. by ψ^*—in general, ψ is a complex function), carrying out the differentiation of the function ψ standing on the right of the operator $\hat{p}$, and integrating over all coordinates:

$$(2m)^{-1} \int_{-\infty}^{\infty} \{[\psi^*(\hat{p}^2\psi)] - [Ze^2/(4\pi\epsilon_0\epsilon)]\psi^*\psi r^{-1}\} \,\mathrm{d}x\, \mathrm{d}y\, \mathrm{d}z = E_n. \tag{A1.4}$$

For example, the lowest-lying energy level ($n = 1$) at -13.6 eV (i.e. we need an energy of 13.6 electron-volts to take an electron from the ground state of a hydrogen atom to a point at an infinitely large distance from the nucleus) has a wave function (solution of equation A1.2 with $Z = 1$, $\epsilon = 1$)

$$\psi(r) = (\pi a_0^3)^{-1/2} \exp(-r/a_0); \qquad a_0 \equiv a = 0.529 \times 10^{-10}\,\mathrm{m}. \tag{A1.5}$$

a_0 is the Bohr radius. We can see that this function is totally symmetric (it depends only on the radial distance from the nucleus). The solutions with higher n decay more slowly (the orbital radius is na_0) and ψ exhibits $n - 1$ nodes near the origin.

Appendix 2

Particle in a one-dimensional rectangular well

Consider a well of width a and depth V. If the barrier seen by a particle in the well is very high ($V \to \infty$), the amplitude of the wave function must be zero at the well boundary ($x = 0$ and $x = a$). Inside the well, the potential is a constant and the solution of the Schrödinger equation is of the form

$$\phi = A \exp(ikx) + B \exp(-ikx), \tag{A2.1}$$

where A and B are determined from the boundary condition ($\phi(0) = \phi(a) = 0$). Hence

$$\phi = \text{const.} \sin\left(\frac{\pi l x}{a}\right), \qquad l = 1, 2, \ldots . \tag{A2.2}$$

The energy is obtained from

$$\left\{-\frac{\hbar^2}{2m}\frac{d^2}{dx^2} + V\right\}\phi = E\phi. \tag{A2.3}$$

Substitute for ϕ from eqn (A2.2), multiply the equation from the left by ϕ, and integrate over x to obtain the energy measured from $V = 0$, i.e.,

$$E = \frac{\hbar^2}{2m}\left(\frac{\pi l}{a}\right)^2. \tag{A2.4}$$

Consider now the case when the potential V is finite. In the well, the potential is zero, so that the Schrödinger equation is

$$\left\{\frac{d^2}{dx^2} + k_2^2\right\}\phi = 0, \tag{A2.5}$$

where $k_2 = (2mE/\hbar^2)^{1/2}$. In the barrier, the potential is V, so that

$$\left\{\frac{d^2}{dx^2} - k_1^2\right\}\phi = 0, \tag{A2.6}$$

where $k_1 = [2m(V - E)/\hbar^2]^{1/2}$. When $E < V$, the particle energy is smaller than the barrier height and k_1 is real. Outside the well, the wave function is

$$\phi = A\exp(ik_1x) + B\exp(-ik_1x). \tag{A2.7}$$

Let us label the region on the left I and that on the right III while inside the well is region II. For ϕ to be a decreasing function of x outside the well we must take

$$\phi_{\mathrm{I}} = A\exp(k_1x); \qquad \phi_{\mathrm{III}} = B\exp(-k_1x). \tag{A2.8}$$

In the well, k_2 is real and we have

$$\phi_{\mathrm{II}} = C\sin(k_2x) \qquad \text{and} \qquad \phi_{\mathrm{II}} = D\cos(k_2x). \tag{A2.9}$$

The solutions (A2.8) and (A2.9) and their derivatives must be continuous at the well boundaries at $x = 0$ and $x = a$, so that

$$\phi_{\mathrm{I}}(0) = \phi_{\mathrm{II}}(0); \qquad \phi_{\mathrm{II}}(a) = \phi_{\mathrm{III}}(a);$$

$$\frac{\mathrm{d}\phi_{\mathrm{I}}}{\mathrm{d}x} = \frac{\mathrm{d}\phi_{\mathrm{II}}}{\mathrm{d}x} \quad \text{at} \quad x = 0 \qquad \text{and} \qquad \frac{\mathrm{d}\phi_{\mathrm{II}}}{\mathrm{d}x} = \frac{\mathrm{d}\phi_{\mathrm{III}}}{\mathrm{d}x} \quad \text{at} \quad x = a. \tag{A2.10}$$

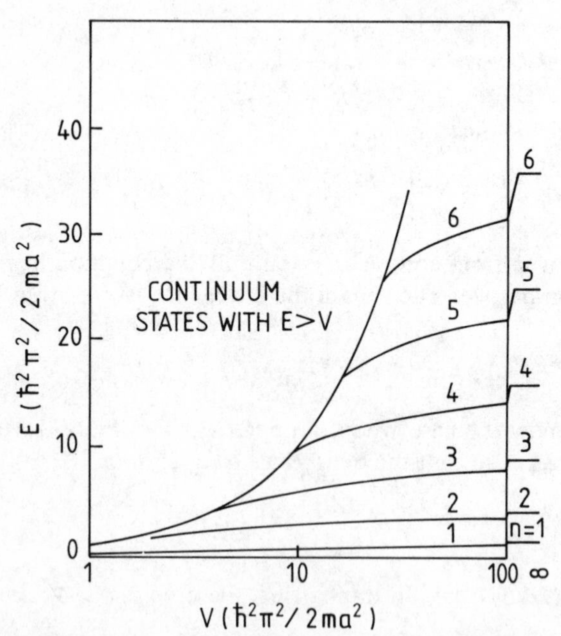

Fig. A2.1. Energy levels of a particle in a one dimensional potential well of depth V and width a.

The energy levels as a function of V and a obtained from eqn (A2.10) are shown in Fig. A2.1. Inside the well, the wave functions resemble the solution of the infinite-barrier problem. However, their amplitude at the well boundaries is finite. Outside the well, these functions decay exponentially.

Appendix 3

Solution of the Schrödinger equation for a periodic one-dimensional (Krönig–Penney) potential

We can solve the Schrödinger equation with the potential shown in Fig. A3.1 along the lines of the argument developed for the particle-in-a-box problem of Appendix 2.

The wave function in the well domain is now defined as

$$\phi_{\mathrm{I}}(x) = A\exp(\mathrm{i}k_1 x) + B\exp(-\mathrm{i}k_1 x); \qquad 0 \leq x \leq a, \qquad \text{(A3.1)}$$

where $\hbar^2 k_1^2/2m = E$. In the barrier domain, we have $(E > V)$

$$\phi_{\mathrm{II}}(x) = C\exp(\mathrm{i}k_2 x) + D\exp(-\mathrm{i}k_2 x); \qquad a < x < a + b, \qquad \text{(A3.2)}$$

where $\hbar^2 k_2^2/2m = E - V$. We require that the solution be a periodic function with period $d = a + b$. This means that $\phi(x) = \phi(x + d)$. For running waves such as $\phi_k(x) = \exp(\mathrm{i}kx)$, we get

$$\phi_k(x + d) = \exp(\mathrm{i}kx)\exp(\mathrm{i}kd) = \phi_k(x)\exp(\mathrm{i}kd).$$

It is possible to prove that any valid solution of the Schrödinger equation with a periodic potential must have this property (the Bloch theorem). We invoked this condition in Chapter 3 to justify the choice of our tight-binding wave function. Hence, we have

$$\phi_{\mathrm{I}}(d) = \phi_{\mathrm{I}}(0)\exp(\mathrm{i}kd) = \phi_{\mathrm{II}}(d). \qquad \text{(A3.3)}$$

With eqn (A3.3), the continuity of the wave function amplitude at $x = 0$ yields

$$A + B = \exp(-\mathrm{i}kd)[C\exp(\mathrm{i}k_2 d) + D\exp(-\mathrm{i}k_2 d)]. \qquad \text{(A3.4)}$$

We also require that the first derivatives satisfy the periodic boundary condition, so that

$$\frac{\mathrm{d}\phi(x + d)}{\mathrm{d}x} = \frac{\mathrm{d}\phi(x)}{\mathrm{d}x}\exp(\mathrm{i}kd). \qquad \text{(A3.5)}$$

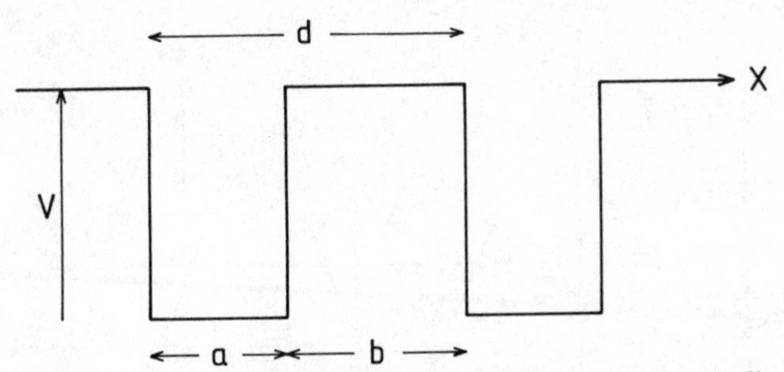

Fig. A3.1. The parameters determining the one-dimensional periodic potential considered in eqn (A3.1).

This gives

$$k_1(A-B)=\exp(-\mathrm{i}kd)\,k_2[C\exp(\mathrm{i}k_2d)-D\exp(-\mathrm{i}k_2d)]. \quad (A3.6)$$

In addition to eqns (A3.4) and (A3.6), which represent the periodicity of the potential, we have the usual continuity conditions for wave function amplitudes and derivatives at $x=a$:

$$A\exp(\mathrm{i}k_1a)+B\exp(-\mathrm{i}k_1a)=C\exp(\mathrm{i}k_2a)+D\exp(-\mathrm{i}k_2a);$$
$$k_1[A\exp(\mathrm{i}k_1a)-B\exp(-\mathrm{i}k_1a)]=k_2[C\exp(\mathrm{i}k_2a)-D\exp(-\mathrm{i}k_2a)]. \quad (A3.7)$$

Thus, we have four homogeneous linear equations for four coefficients A, B, C, and D. Such a set of equations has a non-trivial solution only if the determinant

$$\begin{vmatrix} 1 & 1 & -\exp[\mathrm{i}d(k_2-k)] & -\exp[-\mathrm{i}d(k_2+k)] \\ k_1 & -k_1 & -k_2\exp[\mathrm{i}d(k_2-k)] & k_2\exp[-\mathrm{i}d(k_2+k)] \\ \exp(\mathrm{i}k_1a) & \exp(-k_1a) & -\exp(\mathrm{i}k_2a) & -\exp(-\mathrm{i}k_2a) \\ k_1\exp(\mathrm{i}k_1a) & -k_1\exp(-\mathrm{i}k_1a) & -k_2\exp(\mathrm{i}k_2a) & k_2\exp(-\mathrm{i}k_2a) \end{vmatrix} \quad (A3.8)$$

vanishes. After some algebra we obtain

$$\cos(k_1a)\cos(k_2b)-\sin(k_1a)\sin(k_2b)\left(\frac{k_1^2+k_2^2}{2k_1k_2}\right)=\cos(kd); \qquad E>V. \quad (A3.9)$$

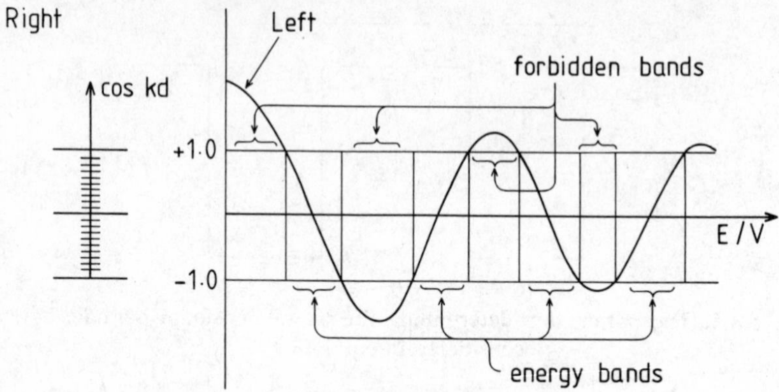

Fig. A3.2. A graphical presentation of the solution shown in eqns (A3.9) and (A3.10), with the allowed and forbidden energy bands clearly indicated.

When $E < V$ we substitute $ik_2 = \mu$, $\hbar^2\mu^2/2m = V - E$, and eqn (A3.9) becomes

$$\cos(k_1 a)\cosh(\mu b) - \sin(k_1 a)\sinh(\mu b)\left(\frac{k_1^2 - \mu^2}{2k_1\mu}\right) = \cos(kd). \quad \text{(A3.10)}$$

The solution is best visualized by plotting the left- and right-hand sides separately as functions of E/V (Fig. A3.2). We can see that the solutions form bands of energies separated by gaps.

Solutions to Problems

1.1. Using the expressions for the Fermi energy and wave vector given in the text and the empirical values of the lattice constant we find:

	a(Å)	$r(a_0)$	E_F(eV)	$\rho(E_F)$ (number of states/$(a_0^3\,\mathrm{eV})$)
Li	3.49	3.25	4.74	2.2 10
Na	4.23	3.93	3.24	1.8 10
K	5.23	4.86	2.12	1.5 10

Note that $\rho(E_F) = mk_F/\hbar^2\pi = 3n/2E_F$, where n is the electron density;

$$a_0 = 0.529 \times 10^{-8}\ \mathrm{cm}; \qquad \text{volume per atom is } 4\pi r^3/3.$$

1.2. In a unit volume there are $(2\pi)^{-2}$ allowed values of k. Then

$$\mathrm{d}n = \frac{2}{(2\pi)^2}\,2\pi k\,\mathrm{d}k; \qquad \rho = \frac{\mathrm{d}n}{\mathrm{d}E};$$

hence, using

$$\frac{\mathrm{d}E}{\mathrm{d}k} = \frac{d}{\mathrm{d}k}\left(\frac{\hbar^2 k^2}{2m}\right),$$

we obtain

$$\rho = \frac{m}{\pi\hbar^2}.$$

1.3.

	Structure	No. of neighbours	a(Å)	Nearest-neighbour distance (Å)
NaCl	FCC	6	5.64	2.82
Si	diamond	4	5.43	2.35
GaAs	zinc-blende	4	5.66	2.45
Na	BCC	8	4.23	3.66

1.4. $E_F(\mathrm{Si}) = 12.45$ eV

1.5. $\int_0^{E_F} \rho(E)\,\mathrm{d}E = n(\text{electron density});$

$E_F = \hbar^2 k_F^2/2m;$ hence

$n = E_F m/\pi\hbar^2 = k_F^2/2\pi$ gives

$k_F = (2\pi n)^{1/2}.$

2.1. $\cos\gamma = (u_1 v_1 + u_2 v_2 + u_3 v_3)(u_1^2 + u_2^2 + u_3^2)^{-1/2}(v_1^2 + v_2^2 + v_3^2)^{-1/2}.$

2.2. The separation of planes in real space is $d = 2\pi/G$, where G is the length of the shortest reciprocal lattice vector, which is

$$(2\pi/a)[h^2 + k^2 + l^2]^{1/2}; \qquad a = 5.43\ \text{Å}.$$

2.3. At the band edge, the wave function is a standing wave (e.g. $\cos[kx]$). The expectation value is proportional to

$$\int \cos(kx)\left(\frac{\mathrm{d}\cos(kx)}{\mathrm{d}x}\right)\mathrm{d}x = 0.$$

2.4. $0.032 = 2\,|V_n| = 2\,|V_1|/100$, so that $|V_1| = 1.6$ (eV).

2.5. $2k_F = 2\pi/a; \quad a = 2$ Å, hence $k_F = \pi/2(\text{Å}^{-1});$

$k_F = (3\pi^2 n)^{1/3},$ hence $n = \pi/3a^3 = (\pi/24) \times 10^{24}\ \mathrm{cm}^{-3}.$

2.6. The band gap is $2 \times 1.6 = 3.2$ (eV), so that the crystal is transparent in the visible range.

3.1.

$$v_g = \frac{\mathrm{d}\omega}{\mathrm{d}k} = \frac{1}{\hbar}\frac{\mathrm{d}E_k}{\mathrm{d}k} = -\frac{2\beta a}{\hbar}\sin(ka).$$

In a metal

$$E_k = \frac{\hbar^2 k^2}{2m} \quad \text{and} \quad v_g = \frac{\hbar k}{m}.$$

3.2.

$$(m^*)^{-1} = \frac{1}{\hbar^2}\frac{\mathrm{d}^2 E_k}{\mathrm{d}k^2} = \frac{2\beta a^2}{\hbar^2}\cos(ka).$$

3.3. If the nearest-neighbour interaction β is the same for all neighbours, then

$$E_{\mathbf{k}} = E_{\mathrm{at}} - 6\beta\cos(\mathbf{k}a).$$

3.4. Going from silicon to GaP and ZnS, we expect bands to narrow, gaps to widen, and charge transfer from cation to anion (ionicity) to increase.

4.1. $$\frac{n_i(\mathrm{Si})}{n_i(\mathrm{Ge})} = \exp\left[\frac{E_g(\mathrm{Ge}) - E_g(\mathrm{Si})}{2k_B T}\right] = \exp(-0.46/0.052) = 1.4 \times 10^{-4}.$$

4.2. All donors are ionized, hence $n = N_d$. Since the intrinsic concentration is

$n_i \approx 10^{10}\ \text{cm}^{-3}$, we have $n \gg n_i$. From eqn (4.6),

$$N_d = n = 12\left(\frac{m_e^* k_B T}{2\pi\hbar^2}\right)^{3/2} \exp\left[-\frac{(E_g - E_F)}{k_B T}\right].$$

The factor 6 arises because there are six equivalent conduction band minima (Chapter 5, Table 4.1). Substitute for N_d, m_e^*, $k_B T$, E_g and take logarithm to evaluate $E_g - E_F = 0.21$ (eV). The Fermi energy therefore lies 0.21 eV above the conduction band edge.

4.3. The force on atom M_1 is proportional to displacement u. We have for the $(2n+1)$th atom an equation of motion

$$M_1 \frac{d^2 u_{2n+1}}{dt^2} = -\alpha(u_{2n+2} - u_{2n+1}) - \alpha(u_{2n} - u_{2n+1}).$$

A similar equation is obtained for a $2n$th atom of mass M_2:

$$M_2 \frac{d^2 u_{2n}}{dt^2} = \alpha(2u_{2n} - u_{2n+1} - u_{2n-1}).$$

Assuming that the motion is harmonic, the displacements are

$$u_n = U_n \exp[i(kx_n - \omega t)].$$

Substituting and rearranging gives two equations:

$$(-\omega^2 M_1 - 2\alpha)U_{2n+1} + 2\alpha \cos(ka) U_{2n} = 0;$$

$$2\alpha \cos(ka) U_{2n+1} + (-\omega^2 M_2 - 2\alpha) U_{2n} = 0.$$

The solution is obtained from the condition that the determinant of the coefficients vanishes:

$$(\omega^2 M_1 + 2\alpha)(\omega^2 M_2 + 2\alpha) - 4\alpha^2 \cos^2(ka) = 0.$$

Solve for ω^2 and plot $\omega(k)$ to obtain the result in Fig. 4.4b.

4.4. Since the momentum carried by our photon is negligible, the only transitions satisfying the momentum conservation rule are the vertical transitions. In the nearly-free-electron model, the only significant contribution comes from the standing waves from the Brillouin zone boundary. According to eqn (4.11), the transition probability there is

$$\sim \left| \int \cos(kx) x \sin(kx)\, dx \right|^2.$$

Elsewhere the probability is proportional to $|\int e^{-ikx} x e^{ik'x}\, dx|^2$ $(k' = k + G)$.

4.5. The binding energy is $m^* Z^2/\epsilon^2 m$ in units of Ry (i.e. 13.6 eV). In the case of interest $Z = 2$. We get from Table 4.1

$$\Delta E_1(\text{Se}^+) = 125\ \text{meV}; \qquad \Delta E_1(\text{Zn}^-) = 208\ \text{meV}.$$

4.6. At room temperature, all donors and acceptors are ionized. Using eqn (4.16) with $n_n/n_p \approx N_d N_a/n_i^2$, we obtain

$$n_i(\text{Si}) = 1.5 \times 10^{10}\ \text{cm}^{-3}; \qquad n_i(\text{GaAs}) = 1.8 \times 10^{6}\ \text{cm}^{-3}.$$

The contact potential $e\phi(\text{Si}) = 0.47(\text{eV})$ and $e\phi(\text{GaAs}) = 0.59(\text{eV})$.

4.7. From the charge neutrality condition, we have

$$l_n/l_p = N_a/N_d = 10^5,$$

so that $l_n \gg l_p$ and $l \approx l_n$. Substitute into eqns (4.25) and (4.26) to obtain $l = 3.3\ \mu\text{m}$.

4.8. Use eqn (4.32) and Table 4.1 to get $D = 39\ \text{cm}^2\ \text{s}^{-1}$:

$$j_{\text{diff}} = 4.16(\text{A cm}^{-2})$$

Set $j_{\text{diff}} = j_d = en\mu_e E$. The electric field E is therefore $3.5 \times 10^2(\text{V cm}^{-1})$.

4.9. Conductivity $\sigma = ne^2\tau/m^*$. If τ and m^* are the same, σ differs because of the difference in electron density: $n(\text{semiconductor}) \simeq N_d < 10^{19}\ \text{cm}^{-3}$, and $n(\text{metal}) \simeq 10^{24}\ \text{cm}^{-3}$.

4.10. Using Table 4.1, we obtain $E_g(\text{Si}) = 1.12\ \text{eV}$. Hence,

$$\frac{hc}{\lambda_{\text{threshold}}} = 1.18\ (\text{eV}); \qquad \lambda_{\text{threshold}} = 10535(\text{Å});$$

$$\Delta n = \Delta n_0 \exp(-t/\tau);$$

$$t = \tau \ln(\Delta n/\Delta n_0) = 10^{-6} \ln(0.1) = 2.3\ \mu\text{s}.$$

5.1. (a) Nearest-neighbour distance is $a\sqrt{3}/4$; next-nearest-neighbour distance is $a/\sqrt{2}$.
(b) $a(\text{Si})$ and $a(\text{GaP})$ are nearly identical; the same is true for other materials from the same row.

5.2.
$$E_F = \left(\frac{\hbar^2}{2m}\right)\left(\frac{3\pi^2 32}{a^3}\right)^{2/3}; \quad E_F(\text{GaAs}) = 11.5\ \text{eV}.$$

5.3. Hint: examine Figs. 2.4 and 5.2.

5.4. For example, GaAs is more ionic than Ge and ZnSe is more ionic than GaAs.

5.5. For example, the conduction band minima near the X and L points. The optical transitions across the (vertical) gap at these points will be strong.

5.6. The fundamental gap is hc/λ; $\lambda \simeq 6200$ Å gives an energy of about 2 eV. We need, for instance, a $Zn_{0.5}Cd_{0.5}Te$ alloy. Neither Si nor GaAs would do.

5.7. $x \simeq 0.54$; $a \simeq 5.54$ Å.

5.8. Hint: recall that the impurity binding energy is the effective Rydberg, i.e. $13.6(m^*/m)(Z/\epsilon)^2(\text{eV})$. Compare this energy with $k_B T$ and use the Boltzmann exponential law to evaluate the probability of ionization.

6.1. 33 Å.

6.2. 2.25 Å cm^{-2}.

6.3. 860 Å.

6.4. $C_{ox} = 3.5 \times 10^{-8}\ \text{F cm}^{-2}$; $l = 2.2 \times 10^{-5}$ cm.

$-eN_a l/C_{ox} = -2\text{V}$.

$e\phi = k_B T \ln(N_a/n_i) = 0.37$ eV.

$V_T = -0.94 - (-2) + 0.74 = 1.8$ eV.

6.5. (a) From Fig. 5.5 the difference in the band gap at Γ at room temperature of the two semiconductors is 0.4 eV. Hence, for the conduction band minima to be level, the band structure of the alloy must be shifted down by 0.4 eV.
(b) The Γ–*X* gap of the alloy is about 40 meV larger. Hence the band structure of the alloy must be shifted down by 40 meV.

6.6. 47.5% of Ga and 52.5% of In; $\Delta E_v = 0.24$ eV with the InP valence band lying lower; $\Delta E_c = 0.2$ eV (T = 0K).

6.7. The width $l = 424$ Å ($\Delta E_c = 0.25$ eV).

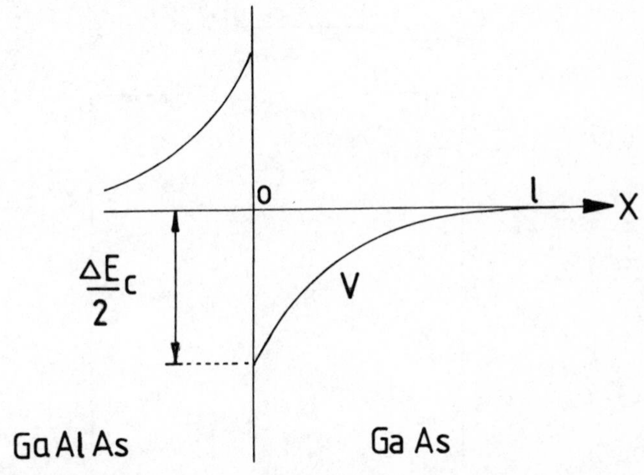

6.8. From eqn (6.7) the space charge potential energy is $V = -e^2 N_d (l-x)^2/2\epsilon_0\epsilon$, $0 \leqslant x \leqslant l$; $V = 0$, $x > l$, where $x = 0$ is the position of the interface.

Energy $E = E(\text{kinetic}) + E(\text{potential})$:

$$E = \int_0^\infty \psi\{-\hbar^2/2m^*\}\{\mathrm{d}^2/\mathrm{d}x^2\}\psi\,\mathrm{d}x + \int_0^l \psi V \psi\,\mathrm{d}x,$$

where $\psi = (b^3/2)^{1/2} \times \exp(-bx/2)$.

$$E = \hbar^2 b^2/8m^* - 0.125(\text{eV})\{1 - 6/bl + 12/b^2 l^2\}.$$

From $\mathrm{d}E/\mathrm{d}b = 0$, we obtain (neglecting the term with l^{-2})

$$b \cong \{5m^* e/l\hbar^2\}^{1/3} \quad (m^{-1}); \qquad \frac{1}{b} = 26\ \text{Å}.$$

6.9. Taking $\psi^2 \sim x^2 \exp(-bx)$, we find $d(\psi^2)/dx = 0$ at $x = 2/b \cong 52$ Å. The peak of the probability density occurs 52 Å from the interface.

6.10. From Table 5.1 and eqn (6.3), we obtain $\Delta E_c \simeq E_g^f(\text{GaP}) - E_g^f(\text{Si}) - \Delta E_v = 2.38 - 1.13 - 0.41 = 0.84$ (eV).

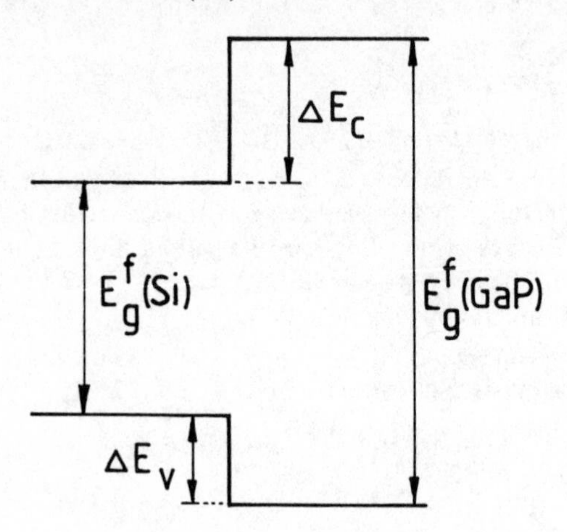

6.11.

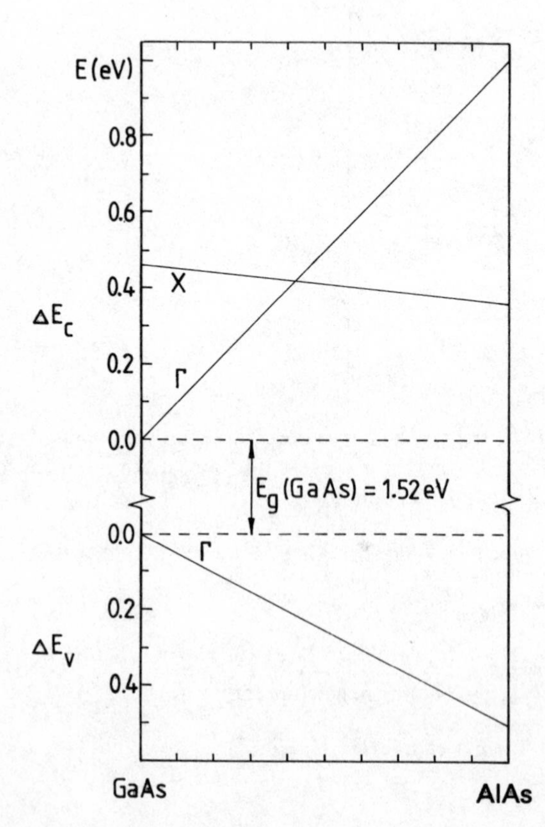

7.1. $\Delta E_c = [E_g(\mathrm{Ga_{0.7}Al_{0.3}As}) - E_g(\mathrm{GaAs})] - \Delta E_v \simeq 0.4 - 0.15 = 0.25$ (eV). $\Delta E_v = 30\%$ of ΔE_v quoted in Table 6.3 for AlAs–GaAs $= 0.5 \times 0.3 = 0.15$ (eV).

$$m_e^*/m = 0.067; \qquad m_h^*/m = 0.62.$$

When $a = 200$ Å, a good estimate can be obtained assuming that the barrier is infinitely high. Then $E_1 = \hbar^2(\pi/a)^2/2m^*$,

$$E_{1e} = 0.014 \text{ ev}; \qquad E_{1hh} = 0.0015 \text{ eV}.$$

When $a \to 0$ the levels are pushed towards the top of the barrier. Hence the difference in the magnitude of the forbidden gap ranges from 15.5 meV to just under 400 meV. A more accurate solution can be obtained from results summarized in Fig. A2.1.

7.2. Using $n_i(\mathrm{GaAs}) \simeq 2 \times 10^6\ \mathrm{cm^{-3}}$, we obtain 0.615 eV.

7.3. Since the band width $\delta E = 100$ meV is large, we can use the nearly-free-electron model to estimate the miniband width and ignore the magnitude of the gaps. Setting $\delta E = \hbar^2 k_{\mathrm{BZ}}^2/2m_e^*$, where $k_{\mathrm{BZ}} = \pi/2a$, we find $a \simeq 19$ Å.

7.4. The width of the depletion layer $l = d/2$, where d is the superlattice period. Since the potential (V) varies as l^2, V decreases with decreasing l.

7.5. $V_0 = esa^2/2\epsilon_0\epsilon$. If $8eV_0 = E_g$, then $s = 2 \times 10^{18}\ \mathrm{cm^{-3}}$.

7.6.
$$E_n = \hbar\omega(n + \tfrac{1}{2}); \qquad V = \tfrac{1}{2}qz^2 = \frac{1}{2}\frac{e^2 s}{\epsilon_0\epsilon} z^2; \qquad s = 2 \times 10^{18}\ \mathrm{cm^{-3}};$$

$$\omega = (q/m^*)^{1/2};$$

$$E_0 = \frac{\hbar}{2}\left(\frac{e^2 s}{\epsilon_0 \epsilon m^*}\right)^{1/2}; \qquad E_0^c = 13 \text{ meV}; \qquad E_0^v = 0.1 \text{ eV};$$

$$E_g(\text{nipi}) = E_g^f(\mathrm{Si}) - 2V_0 + |E_0^c| + |E_0^v|;$$

$$E_g^f - E_g(\text{nipi}) \cong 0.26 \text{ eV}.$$

7.7. Since the barrier height $\Delta E_v = 0.5$ eV is large, we can simplify our calculation by using the infinite-barrier result. (Otherwise we must use more accurate results presented in Fig. A2.1.) Then,

$$E_1 \cong 24 \text{ meV}; \qquad E_1(\text{wire}) \cong 48 \text{ meV}; \qquad E_1(\text{box}) \cong 72 \text{ meV}.$$

7.8. ($\mathrm{Ga_{0.7}Al_{0.3}As}$) $\simeq 1.82$ eV (Fig. 5.5) hence $\lambda \simeq 6500$ Å.

7.9. For blue light, $\lambda \simeq 4700$ Å. We can use an alloy of ZnSeTe or an equivalent quantum well structure.

7.10. For example, α–Sn–InSb.

7.11. An alloy of $\mathrm{GaAs_{0.7}P_{0.3}}$; $E_g \simeq 1.895$ eV (at room temperature only about 1.8 eV) and hence $\lambda \simeq 6556$ Å (at 300 K, $\lambda \simeq 6900$ Å). Using a 30 Å quantum well of GaAs and taking $m_e^*/m = 0.1$ and $m_{hh}^*/m = 0.6$ gives $E_g \simeq 2.01$ eV and $\lambda \simeq 6181$ Å; a 50 Å well gives $\lambda \simeq 7330$ Å.

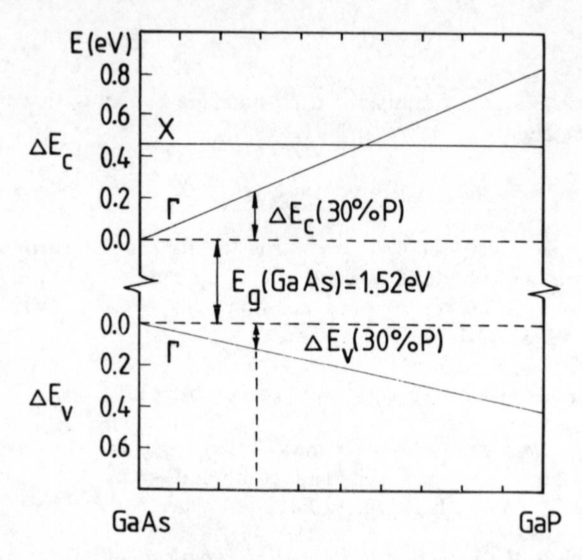

8.1. $\Delta E = 0.25$ eV; $m_e^*/m = 0.067$; $T^{-1} = 64$.

8.2. Using the triangular barrier formula, we obtain $T = 0.04$.

8.3. The width of the space charge layer is 0.07 μm. The potential barrier at the junction is 1.06 eV and the maximum electric field is 2×10^6 V cm^{-1}.

8.4. The electric field is 640 V cm^{-1}.

8.5. The electric field is 10 V cm^{-1}.

8.6. The minimum energy needed equals the magnitude of the fundamental gap (e.g. in GaAs 1.41 eV at room temperature).

It takes about 0.1 ps to emit a phonon; hence, it would take about 14 ps for the electron to relax to the bottom of the conduction band.

8.7. The electric field is 1.26×10^4 V cm^{-1}.

8.8. The thermionic emission current density is $j = 120\,(\text{A cm}^{-2}\,\text{K}^{-2})\; T^2 \exp(-W/k_BT) = 5 \times 10^3$ A cm^{-2}. The tunnelling current is $|j_t| = env$ times the transmission coefficient:

$$v = (3k_BT/m_e^*)^{1/2} = 4.55 \times 10^7 \text{ cm s}^{-1};$$

$$n = 2(2\pi m_e^* k_BT/h^2)^{3/2} \exp[-(E_c - E_F)/k_BT]; \qquad -(E_c - E_F) = 50 \text{ meV};$$

$$n \cong 3 \times 10^{18} \text{ cm}^{-3}.$$

Hence, $|j_t| = 9 \times 10^5$ A cm^{-2}.

8.9. The space charge width in the p-type region is very small and can be neglected. Hence,

$$l = l_n = \left[\frac{2\epsilon_0 \epsilon k_B T}{e^2 N_d} \ln\left(\frac{N_a N_d}{n_i^2}\right)\right]^{1/2} \simeq 3.3 \times 10^{-4} \text{ cm}.$$

The density of holes in the p-region is $p_p = 5 \times 10^{19}$ cm^{-3}, and in the n-region $p_n = n_i^2/N_d = (1.5 \times 10^{10})^2/10^{14} = 2.25 \times 10^6$ cm^{-3}. The gradient dp/dx responsible for hole diffusion equals $(p_p - p_n)/l$. Hence $dp/dx = 1.5 \times 10^{23}$ cm^{-4} and the diffusion current density is $eD(dp/dx) = 3 \times 10^5$ A cm^{-2}.

In equilibrium (and in the absence of an external field) the diffusion current must be equal and opposite to the drift current.

9.1.
$$\frac{\chi^{(1)}(\text{Si})}{\chi^{(1)}(\text{H})} \simeq \left(\frac{E_{2p} - E_{1s}}{E_g}\right)^2 = (\tfrac{3}{4} \times 13.6/5)^2 = 4.2.$$

9.2. $\omega_0 \simeq 0$ for metals; $\omega_0 = E_g/\hbar$ for silicon. Hence, when $\omega \to 0$, $\chi^{(1)}(\text{metal}) \simeq \omega^{-2} \to \infty$, whereas $\chi^{(1)}(\text{Si}) \to (\hbar\omega_p/E_g)^2 = \text{const}$.

9.3. $\omega \simeq \text{mass}^{-1/2}$; $m(\text{atom})/m(\text{electron})$ is of order 10^4–10^5; hence,

$$\frac{\omega(\text{el})}{\omega(\text{vib})} \simeq 100\text{–}1000;$$

$\omega(\text{vib})$ is in the infrared range.

9.4. Taking $n = \epsilon^{1/2}(\gg 1) \simeq [\omega_p^2/(\omega_0^2 - \omega^2)]^{1/2}$, where $\omega_0 = E_g/\hbar$, we evaluate $\Delta k \simeq 0.12\omega_p/2c$ and $l_c = 2\pi/\Delta k \simeq 1.3$ μm.

9.5. $\chi_s/\chi_b \simeq 50$ for GaAs–$Ga_{0.7}Al_{0.3}As$ and 1.6 for InSb–CdTe.

9.6. $\chi_s/\chi_b \sim m^*$ and $m_{hh}^*/m_e^* = 9.25$ (GaAs) and 34.3 (InSb).

9.7. The absorption must be weak enough to avoid saturation. At high temperatures, the population n_2 of higher miniband 2 can be significant because the separation of minibands δE is typically 10–100 meV, i.e., $n_2/n_1 \simeq \exp[-(\delta E/k_B T)]$.

9.8. Since eqn (9.70) is based on the tight-binding model, it is only valid provided that the wells are separated by thick barriers. Then the overlap between wave functions associated with adjacent wells—and consequently the width of the miniband—are small.

9.9. We know that the susceptibility is approximately given by

$$\chi_s = \chi^{(1)} m^* \Delta \, d^2 \cos(kd)/\hbar^2.$$

At $k = 0$, $\cos(0) = 1$; and at $\delta k > 0$, $\cos(\delta kd) > 0$; at $k = \pi/2d$, $\cos(\pi/2) = 0$; and at $k = \pi/2d + \delta k$, $\cos(kd) < 0$.

9.10. A valence electron can only be excited across the band gap, so that $\omega_0 \simeq E_g/\hbar$. An electron placed at the conduction band can be excited to the nearest miniband at an energy δE above. Since the susceptibility is inversely proportional to ω_0^2, the ratio of the valence (χ_v) and conduction (χ_c) electron contributions is $\chi_v/\chi_c \simeq (\delta E/E_g)^2$.

10.1. The nearest separation in silicon is $a/\sqrt{2}$ ($a = 5.43$ Å); in NaCl it is $a/2$ ($a = 5.64$ Å).

10.2. E(NaCl) = 18.9 eV; E(Si) = 10.2 eV; λ is given in eqn (10.1).

10.3. Kinetic energy = $80 - 20 - 7 = 53$ (eV) $= \frac{1}{2}mv^2$; $v = 4.3 \times 10^8$ cm s^{-1}.

10.4. For the resonance to be observable,

$$\omega_c\tau = eB\tau/m^* > 1,$$

hence $B > m^*/e\tau$, where τ is the relaxation time.

The mean free path is approximately $l = N_d A = \tau v_{\text{th}} = \tau\ (3k_BT/m^*)^{1/2}$, $N_d = 10^{16}$ cm^{-3} and $A = \pi r^2$, where r is the radius of the impurity orbit in germanium. For the thermal energy, we have taken the classical value for a free particle with three degrees of freedom, i.e. $3k_BT/2$. Substituting, we obtain $B > 0.03$ tesla.

10.5. Use eqn (10.6). From the data, $E_{e-hh} \simeq 1.895$ eV at crossover.

$$E_g^X(\text{Ga}_{0.72}\text{Al}_{0.28}\text{As}) = 0.72E_g^X(\text{GaAs}) + 0.28E_g^X(\text{AlAs}) = 2.09 \text{ eV}.$$

(see Table 5.2). $E_{\text{ex}} \simeq 9$ meV and $\Delta_{hh} \simeq (\hbar^2/2m_{hh}^*)(\pi/d)^2 = 12$ meV, where $d = 68$ Å and $m_{hh}^*/m = 0.62$. Hence, $\Delta E_v = 2.09 + 0.009 - 1.895 - 0.013 = 0.191$ (eV).

11.1. $E_c - E_F \simeq 0.11$ eV; sheet electron density is $N_s \simeq 2 \times 10^{12}$ cm^{-2}; $E_1 \simeq 0.24$ eV.

11.2. The transit time is proportional to mobility. Since the electron effective mass is about five times smaller in GaAs, mobility is larger in the HEMT. An additional enhancement of mobility takes place at low temperatures owing to the effect of modulation doping, as shown in Fig. 8.14.

11.3. $T = 0.34$.

11.4. Using the thermal velocity, the transit time is about 150 fs. If there are no collisions, and the field acts on the electron along the full length of the base, the transit time is shortened to about 20 fs.

11.5. $a = 23$ Å.

12.1. For a cavity of width L, the wavelength λ at resonance is (where m labels the modes and n is the refractive index) $m\lambda = 2nL$, so that $\mathrm{d}m/\mathrm{d}\lambda = -2Ln/\lambda^2 + (2L/\lambda)\mathrm{d}n/\mathrm{d}\lambda$. Since the last term is small, we obtain, for $\mathrm{d}m = 1$, $\mathrm{d}\lambda = 2.24$ Å.

12.2. The gain is proportional to the population inversion density, which is in turn proportional to the density of states in the region of energies $\hbar\omega$ just above the bulk band gap energy of GaAs, E_g (Fig. 12.1). Hence, gain $\sim(\hbar\omega - E_g)^{1/2}$ or $(\omega - E/\hbar)^{1/2}$.

12.3. Assuming that all loss factors scale with the refractive index n, we have

$$R'(Ga_{0.62}Al_{0.38}As) = [(3.464 - 1)/(3.464 + 1)]^2 = 0.304; \qquad R = 0.323;$$

$$n^2(Ga_{0.62}Al_{0.38}As) = \epsilon(Ga_{0.62}Al_{0.38}As) = 0.62\epsilon(GaAs) + 0.38\epsilon(AlAs) = 12.$$

Hence $\ln R'/\ln R = 1.05$. A 5% increase in gain amounts to about 0.5% increase in the threshold current (Fig. 12.2).

12.4. Vibrational frequency $\omega \sim (M_1^{-1} + M_2^{-1})^{1/2}$. Hence $\omega(SiO_2)/\omega(KCl) = 2.24$: the wavelength is 2.24 times larger in KCl.

12.5. The binding energy of excitons increases as a result of increased confinement. Hence the exciton is more stable at higher temperatures.

12.6. (a) Use the definition of polarizability as an expansion

$$P = \sum_j \epsilon_0 \chi^{(j)} \mathscr{E}_0^j.$$

When the external field $\mathscr{E}$ is so strong that it can compete with the internal crystalline field, the adjacent terms in the series will be comparable, i.e. $\chi^{(3)}\mathscr{E}_0^3 \approx \chi^{(1)}\mathscr{E}_0$. Since the internal fields must be of order E_g/e per angstrom, we obtain $\chi^{(3)}/\chi^{(1)} \approx 10^{-16}$ $(\mathrm{V\,cm^{-1}})^2$.
(b) $n = n_0 + n_2[\mathrm{SI}]I(\mathrm{W\,m^{-2}})$ and $n_2[\mathrm{SI}] = (4\pi/3) \times 10^{-7}\, n_2[\mathrm{CGS}]/n_0$.

The periodic table

THE PERIODIC TABLE

The letters s, p, d,... signify electrons having orbital angular momentum 0,1, 2,...in units $\hbar$; the number to the left of the letter denotes the principal quantum number of one orbit; and the superscript to the right denotes the number of eletrons in the orbit.

H^{1} 1s																	He^{2} $1s^2$
Li^{3} 2s	Be^{4} $2s^2$											B^{5} $2s^22p$	C^{6} $2s^22p^2$	N^{7} $2s^22p^3$	O^{8} $2s^22p^4$	F^{9} $2s^22p^5$	Ne^{10} $2s^22p^6$
Na^{11} 3s	Mg^{12} $3s^2$											Al^{13} $3s^23p$	Si^{14} $3s^23p^2$	P^{15} $3s^23p^3$	S^{16} $3s^23p^4$	Cl^{17} $3s^23p^5$	Ar^{18} $3s^23p^6$
K^{19} 4s	Ca^{20} $4s^2$	Sc^{21} 3d $4s^2$	Ti^{22} $3d^2$ $4s^2$	V^{23} $3d^3$ $4s^2$	Cr^{24} $3d^5$ 4s	Mn^{25} $3d^5$ $4s^2$	Fe^{26} $3d^6$ $4s^2$	Co^{27} $3d^7$ $4s^2$	Ni^{28} $3d^8$ $4s^2$	Cu^{29} $3d^{10}$ 4s	Zn^{30} $3d^{10}$ $4s^2$	Ga^{31} $4s^24p$	Ge^{32} $4s^24p^2$	As^{33} $4s^24p^3$	Se^{34} $4s^24p^4$	Br^{35} $4s^24p^5$	Kr^{36} $4s^24p^6$
Rb^{37} 5s	Sr^{38} $5s^2$	Y^{39} 4d $5s^2$	Zr^{40} $4d^2$ $5s^2$	Nb^{41} $4d^4$ 5s	Mo^{42} $4d^5$ 5s	Tc^{43} $4d^6$ 5s	Ru^{44} $4d^7$ 5s	Rh^{45} $4d^8$ 5s	Pd^{46} $4d^{10}$ –	Ag^{47} $4d^{10}$ 5s	Cd^{48} $4d^{10}$ $5s^2$	In^{49} $5s^25p$	Sn^{50} $5s^25p^2$	Sb^{51} $5s^25p^3$	Te^{52} $5s^25p^4$	I^{53} $5s^25p^5$	Xe^{54} $5s^25p^6$
Cs^{55} 6s	Ba^{56} $6s^2$	La^{57} 5d $6s^2$	Hf^{72} $4f^{14}$ $5d^2$ $6s^2$	Ta^{73} $5d^3$ $6s^2$	W^{74} $5d^4$ $6s^2$	Re^{75} $5d^5$ $6s^2$	Os^{76} $5d^6$ $6s^2$	Ir^{77} $5d^9$ –	Pt^{78} $5d^9$ 6s	Au^{79} $5d^{10}$ 6s	Hg^{80} $5d^{10}$ $6s^2$	Ti^{81} $6s^26p$	Pb^{82} $6s^26p^2$	Bi^{83} $6s^26p^3$	Po^{84} $6s^26p^4$	At^{85} $6s^26p^5$	Rn^{86} $6s^26p^6$
Fr^{87} 7s	Ra^{88} $7s^2$	Ac^{89} 6d $7s^2$															

Ce^{58} $4f^2$ $6s^2$	Pr^{59} $4f^3$ $6s^2$	Nd^{60} $4f^4$ $6s^2$	Pm^{61} $4f^5$ $6s^2$	Sm^{62} $4f^6$ $6s^2$	Eu^{63} $4f^7$ $6s^2$	Gd^{64} $4f^7$ 5d $6s^2$	Tb^{65} $4f^8$ 5d $6s^2$	Dy^{66} $4f^{10}$ $6s^2$	Ho^{67} $4f^{11}$ $6s^2$	Er^{68} $4f^{12}$ $6s^2$	Tm^{69} $4f^{13}$ $6s^2$	Yb^{70} $4f^{14}$ $6s^2$	Lu^{71} $4f^{14}$ 5d $6s^2$
Th^{90} – $6d^2$ $7s^2$	Pa^{91} $5f^2$ 6d $7s^2$	U^{92} $5f^3$ 6d $7s^2$	Np^{93} $5f^4$ $7s^2$	Pu^{94} $5f^6$ $7s^2$	Am^{95} $5f^7$ $7s^2$	Cm^{95} $5f^7$ 6d $7s^2$	Bk^{97}	Cf^{98}	Es^{99}	Fm^{100}	Md^{101}	No^{102}	Lw^{103}

Fundamental constants

Quantity	Value/units	CGS	SI
Electron charge (e)	1.60219	$\times$	10^{-19}C
	4.80324	$\times 10^{-10}$ e.s.u.	—
Electron-volt (eV)	1.60219	$\times 10^{-12}$ erg eV^{-1}	10^{-19} J eV^{-1}
Electron mass (m)	9.1095	$\times 10^{-28}$ cm	10^{-31} kg
Planck's constant (h)	6.6262	$\times 10^{-27}$ erg s	10^{-34} J s
Planck's constant ($\hbar$)	1.05459	$\times 10^{-27}$ erg s	10^{-34} J s
Bohr radius (a_0)	0.529177	$\times 10^{-8}$ cm	10^{-10} m
Rydberg (Ry)	13.6058	$\times$ 1 eV	1 eV
Speed of light (c)	2.997925	$\times 10^{10}$ cm s^{-1}	10^{8} m s^{-1}
Avogadro's constant (N_A)	6.022	$\times 10^{23}$ mol^{-1}	10^{23} mol^{-1}
Boltzmann's constant (k_B)	1.3807	$\times 10^{-16}$ erg K^{-1}	10^{-23} J K^{-1}
Boltzmann's constant (k_B)	8.617	$\times 10^{-5}$ eV K^{-1}	10^{-5} eV K^{-1}
Permittivity of free space (ϵ_0)	8.854	$\times$	10^{-12} F m^{-1}
Permeability of free space ($\mu_0 = 1/\epsilon_0 c^2$)	4π	$\times$	10^{-7} H m^{-1}
Energy $k_B T (T = 273.15\ \mathrm{K})$	2.3538	$\times 10^{-2}$ eV	10^{-2} eV
Proton mass	1.6726	$\times 10^{-24}$ g	10^{-27} kg
Proton–electron mass ratio	1836.15	$\times$ 1	1
Bohr magneton ($\mu_B = e\hbar/2m$)	9.274	$\times 10^{-21}$ erg G^{-1}	10^{-24} J T^{-1}

Bibliography

General solid-state physics

The reader is assumed to be familiar with solid-state and semiconductor physics at the level of, say, M. N. Rudden and J. Wilson's *Elements of solid state physics* (Wiley, New York, 1980). A more detailed account at a somewhat more advanced level of the material presented in Chapters 1–5, and of many other topics, can be found, for example, in C. Kittel's *Introduction to solid state physics* (Wiley, New York, 1971), in N. W. Ashcroft and N. D. Mermin's excellent *Solid state physics* (Holt, Rinehart and Winston, New York, 1976), or in the most recent book by G. Burns, *Solid state physics* (Academic Press, New York, 1985), which has the broadest scope and may serve as a reference book.

Further insights into key aspects of semiconductor physics can be obtained from the following specialist monographs.

Harrison, W. A. (1980). *Electronic structure and properties of solids*. Freeman, San Francisco.

Jaros, M. (1982). *Deep levels in semiconductors*. Adam Hilger, Bristol.

Nag, B. R. (1980). *Electron transport in compound semiconductors*. Springer, Berlin.

Pearsall, T. P. (ed.), (1982). *GaInAsP alloy semiconductors*. Wiley, New York.

Phillips, J. C. (1973). *Bonds and bands in semiconductors*. Academic Press, New York.

Ridley, B. K. (1982). *Quantum processes in semiconductors*. Clarendon Press, Oxford.

Quantum mechanics

Any elementary textbook on quantum mechanics covers the background knowledge assumed here quite adequately. Of the books at an intermediate level of difficulty suitable for further reading, I suggest R. L. Liboff's *Quantum mechanics* (Holden-Day, San Francisco). An advanced presentation can be found, for example, in an excellent text by E. Mertzbacher, *Quantum mechanics* (Wiley, New York, 1970).

Semiconductor devices

Only elementary knowledge of solid-state electronics, e.g. at the level of Rudden and Wilson's textbook, is assumed here. More details concerning standard topics in this field can be found, for instance, in D. A. Fraser's *The physics of semiconductor devices* (Oxford University Press, Oxford, 1979), or in J. E. Carrol's *Physical models for semiconductor devices* (Arnold, London, 1980).

Optoelectronics

Topics peculiar to applications in optoelectronics are covered in elementary but quite detailed textbooks by J. Wilson and J. F. Hawkes, *Optoelectronics: an introduction* (Prentice-Hall, London, 1983) and by J. M. Senior, *Optical fiber communications* (Prentice-Hall, London, 1985). Many advanced texts are also available, for instance:

Bergh, A. A. and Dean, P. J. (1976). *Light emitting diodes.* Clarendon Press, Oxford.

Casey, H. C. and Panish, M. B. (1978). *Heterostructure lasers.* Academic Press, New York.

Haus, H. A. (1984). *Waves and fields in optoelectronics.* Prentice-Hall, Englewood Clifts, N.J.

Yariv, A. (1985). *Optical electronics.* Holt-Saunders, New York.

Semiconductor microstructures

A semi-quantitative account of the physics and applications of semiconductor heterojunctions, quantum wells, superlattices, etc., at an elementary or intermediate level such as that adopted in the present book, is not available elsewhere. However, it is worth mentioning that there are numbers of semi-popular articles in, for example, *Scientific American, Physics Today, Nature,* and *New Scientist.* There are also many advanced texts. Indeed, almost every major publishing house runs a series of research monographs, review books, and conference proceedings concerning semiconductor microstructures. For example, a collection of articles by leading researchers, and references to current literature on practically any aspect of this wide field (and certainly on any topic included in the present book), can be found in *The physics and fabrication of microstructures and microdevices,* edited by M. J. Kelly and C. Weisbuch (Springer, Heidelberg, 1986), or in *Heterojunction band discontinuities: physics and device applications,* edited by F. Capasso and G. Margaritondo (North Holland, Amsterdam, 1987).

Publications concerning semiconductor microstructures can be found in almost any journal on solid-state physics, applied physics and solid-state electronics, and material growth and processing. In particular, there is a specialist journal published by Academic Press on *Superlattices and microstructures* in which one finds current conference proceedings and brief review and contributed papers on the subject. In the July/August issues of the *Journal of vacuum science and technology* the reader will find proceedings of the annual Conference on the Physics and Chemistry of Semiconductor interfaces. *Semiconductors and Semime-*

tals has published review articles devoted to topics such as heterojunction lasers and crystal growth, and a forthcoming volume will be devoted to physics and applications of strained layer superlattices and ordered microstructures in general. For those who are interested in the more academic aspects of the physics of two-dimensional systems, many-body effects in a two dimensional electron gas, and the quantum Hall effect in particular, there is a regular International Conference on Electronic Properties of Two Dimensional Systems. The proceedings of this conference series are published in *Surface science*.

Alas! Very often these specialist texts appear quite impenetrable to the general reader, not so much because of advanced conceptual or mathematical content but mainly because their scientific message is obscured by a highly technical style of presentation. In order to assist the reader in bridging the gap between the down-to-earth picture presented here and the material presented in the specialist journals and monographs, I would like to recommend a few review articles and books that are at least in part tutorial in character.

Adachi, S. (1985). GaAs, AlAs and GaAlAs: material parameters for use in research and device applications. *Journal of Applied Physics,* **58,** R1–R29. This is a very useful review of just about anything one might need to know about the parameters required in applications of GaAs and AlAs microstructures, with 174 references to original papers.

Ando, T., Fowler, A. B., and Stern, F. (1982). Electronic properties of two dimensional systems. *Reviews of Modern Physics,* **54,** 437–625. This is a definitive review of the work done in the 1970s, with over 2000 references, and with emphasis on the physics and applications of MOSFET silicon structures.

Bauer, G., Kuchar, F., and Heinrich, H. (eds) (1984). *Two dimensional systems, heterostructures and superlattices,* Springer Series in Solid State Science, Vol. 53, and *Two dimensional systems, Physics and new devices* (Vol. 67, 1986), Springer, New York. These contain lectures from two winter schools on these topics covering a wide range of issues in some detail. This series includes a number of other topics concerning semiconductor microstructures and the reader is certain to find among the titles the topic of his interest.

Chang, L. L. (1980). Molecular beam epitaxy. In *Handbook of semiconductors* (ed. S. P. Keller), p. 563. North Holland, Amsterdam. This presents a general outline of the MBE technology.

Dingle, R. (ed.) (1988). Applications of multiquantum wells, selective doping, and superlattices. *Semiconductors and semimetals,* Vol. 24. Academic Press, New York. This volume is devoted to a detailed account of properties and applications of thin layers of III–V semiconductors and includes both basic concepts and device structures.

Ferry, D. K. (1982). Materials considerations for advances in submicron very large scale integration. *Advances in Electronics and Electron Physics* **58,** 311–390. This addresses electron transport theory and applications.

Hess, K. (1982). Aspects of high field transport in semiconductor heterolayers and semiconductor devices. *Advances in Electronics and Electron Phyiscs* **59,** 230–291. This is concerned with hot-electron transport and with electronic structure effects in particular.

Jaros, M. (1985). Electronic properties of semiconductor alloy systems. *Reports on Progress in Physics,* **48,** 1091–154. This presents an outline of microscopic theory of tunable semiconductor microstructures involving alloys.

Pearsall, T. P. (ed.) (1989). Strained-layer superlattices. *Semiconductors and semimetals,* Academic Press, New York. This is a two-volume survey of crystal growth, experimental and theoretical research as well as practical applications, and embracing all key phenomena peculiar to superlattice structures of any kind.

Ploog, K. and Dohler, G. (1983). Compositional and doping superlattices in III–V semiconductors. *Advances in Physics,* **33,** 285–360. This presents an experimentalist's view of optical properties and applications of superlattices.

Prange, R. E. and Girvin, S. M. (ed.), (1987). The quantum Hall effect, Springer, New York. This gives a broad account of one of the most exciting academic achievements in the field of low-dimensional semiconductors and of magnetotransport in general.

Index